Fire Service Ground Ladders

9th EDITION

Cover And Title Page Photo Courtesy of Michael Wieder

Published by
Fire Protection Publications
Oklahoma State University
Stillwater, Oklahoma

RECYCLABLE

Printed in the United States of America   3   4   5   6   7   8   9   10

Dedication

This manual is dedicated to the members of that unselfish organization
of men and women who hold devotion to duty
above personal risk, who count on sincerity of service above
personal comfort and convenience, who strive unceasingly to find
better ways of protecting the lives, homes and property
of their fellow citizens from the ravages of fire and other
disasters . . . **The Firefighters of All Nations**.

Dear Firefighter:

The International Fire Service Training Association (IFSTA) is an organization that exists for the purpose of serving firefighters' training needs. Fire Protection Publications is the publisher of IFSTA materials. Fire Protection Publications staff members participate in the National Fire Protection Association and the International Association of Fire Chiefs.

If you need additional information concerning our organization or assistance with manual orders, contact:

Customer Services
Fire Protection Publications
Oklahoma State University
930 N. Willis
Stillwater, OK 74078-8045
1 (800) 654-4055

For assistance with training materials, recommended material for inclusion in a manual, or questions on manual content, contact:

Editorial Department
Fire Protection Publications
Oklahoma State University
930 N. Willis
Stillwater, OK 74078-8045
(405) 744-5723•Fax (405) 744-4112•Email: editors@osufpp.org

THE INTERNATIONAL FIRE SERVICE TRAINING ASSOCIATION

The International Fire Service Training Association (IFSTA) was established as a "nonprofit educational association of fire fighting personnel who are dedicated to upgrading fire fighting techniques and safety through training." This training association was formed in November 1934, when the Western Actuarial Bureau sponsored a conference in Kansas City, Missouri. The meeting was held to determine how all the agencies interested in publishing fire service training material could coordinate their efforts. Four states were represented at this initial conference. Because the representatives from Oklahoma had done some pioneering in fire training manual development, it was decided that other interested states should join forces with them. This merger made it possible to develop training materials broader in scope than those published by individual agencies. This merger further made possible a reduction in publication costs, because it enabled each state or agency to benefit from the economy of relatively large printing orders. These savings would not be possible if each individual state or department developed and published its own training material.

To carry out the mission of IFSTA, Fire Protection Publications was established as an entity of Oklahoma State University. Fire Protection Publications' primary function is to publish and disseminate training texts as proposed and validated by IFSTA. As a secondary function, Fire Protection Publications researches, acquires, produces, and markets high-quality learning and teaching aids as consistent with IFSTA's mission. The IFSTA Executive Director is officed at Fire Protection Publications.

IFSTA's purpose is to validate training materials for publication, develop training materials for publication, check proposed rough drafts for errors, add new techniques and developments, and delete obsolete and outmoded methods. This work is carried out at the annual Validation Conference.

The IFSTA Validation Conference is held the second full week in July, at Oklahoma State University or in the vicinity. Fire Protection Publications, the IFSTA publisher, establishes the revision schedule for manuals and introduces new manuscripts. Manual committee members are selected for technical input by Fire Protection Publications and the IFSTA Executive Secretary. Committees meet and work at the conference addressing the current standards of the National Fire Protection Association and other standard-making groups as applicable.

Most of the committee members are affiliated with other international fire protection organizations. The Validation Conference brings together individuals from several related and allied fields, such as:

- Key fire department executives and training officers
- Educators from colleges and universities
- Representatives from governmental agencies
- Delegates of firefighter associations and industrial organizations
- Engineers from the fire insurance industry

Committee members are not paid nor are they reimbursed for their expenses by IFSTA or Fire Protection Publications. They come because of commitment to the fire service and its future through training. Being on a committee is prestigious in the fire service community, and committee members are acknowledged leaders in their fields. This unique feature provides a close relationship between the International Fire Service Training Association and other fire protection agencies, which helps to correlate the efforts of all concerned.

IFSTA manuals are now the official teaching texts of most of the states and provinces of North America. Additionally, numerous U.S. and Canadian government agencies as well as other English-speaking countries have officially accepted the IFSTA manuals.

Table Of Contents

Tables

Preface

This ninth edition of **Fire Service Ground Ladders** is intended to provide firefighters with everything they need to know regarding the design, use, testing, and maintenance of fire service ground ladders. As with any manual we produce, there are a variety of people who are responsible for the final product.

We gratefully thank the members of the IFSTA Ground Ladders validation committee. This manual would not have been possible without their input.

1989 Chair
George Tockstein
Santee, CA

1990 Chair
Ron Cody
Redondo Beach, CA

1989 Secretary
Ron Cody
Redondo Beach, CA

1990 Secretary
Bob Anderson
Spokane, WA

Boyd Cole
Northbrook, IL

Bill Cooper
Huntington Beach, CA

Sam Cramer
Florence, SC

Ron Graw
Rockford, IL

Bob Hasbrook
Branson, MO

Jim Mendonsa
Columbia, CA

Phil Schwab
Oshkosh, WI

Rick Baird
Brampton, Ontario

Dave Clark
Champaign, IL

Kurt Kenworth
Chino, CA

Wes Kitchel
Santa Rosa, CA

Jack Tyler
Vancouver, British Columbia

As this manual is one of the more photo-intensive manuals that IFSTA produces, we relied upon the assistance of various fire departments to complete the project. There were several fire departments, both career and volunteer, whose assistance was invaluable in completing this project.

The Santa Rosa, California Fire Department, under the direction of Fire Chief Tony Pini, was responsible for staging the bulk of the pictures in this manual. Captain Wes Kitchel organized the effort. Special thanks to all the Santa Rosa firefighters, many of whom volunteered to pose on their off-duty days, for their professional assistance and hospitality.

The San Francisco Fire Department, through the assistance of Training Chief Al Da Cuna and Truck 7, also provided valuable assistance in completing the project.

Thanks to the Enid, Oklahoma Fire Department for providing their aerial apparatus and crew for staging some of the ladder handling scenes.

Special thanks to the Tulsa, Oklahoma Fire Department for assisting with some of the last minute items we needed. Visual Communications Coordinator Frank Mason organized the trip. Training Captains Greg Neely and Pat Lemons, Engines 2 and 4, and Ladder 2 helped stage the pictures. It is interesting to note that Captain Neely staged the pompier ladder evolutions for both the eighth and ninth editions of the manual.

Two volunteer fire departments also provided personnel and equipment that are shown in this manual. Thanks to the Collegeville Fire Company No.1 and the Pennsburg Fire Company No.1 of Montgomery County, Pennsylvania for their assistance.

We would also like to extend our thanks to the other people and organizations who provided assistance toward the completion of this manual.

Stillwater (OK) Fire Department
Bill Tompkins, Bergenfield, NJ
Joel Woods, Maryland Fire and Rescue Institute
ALACO Ladder Company, Chino, CA
Phoenix (AZ) Fire Department
Chicago (IL) Fire Department

Gratitude is also extended to the following members of the Fire Protection Publications staff who made contributions toward the final publication of this manual.

Barbara Adams, Senior Publications Specialist
Susan Walker, Coordinator of Instructional Development
Ann Moffat, Graphic Design Analyst
Desa Porter, Senior Graphic Designer
Connie Burris, Senior Graphic Designer
Lori Williamson, Contract Graphic Designer

Lynne C. Murnane
Managing Editor

Introduction

The major objectives of fireground operations are rescue, confinement, and extinguishment. In order to meet these objectives firefighters are required to perform many tasks using a variety of tools and appliances. Ground ladders are among the most basic, yet important, of these tools.

Fire service ground ladders are specially designed, tested, and constructed for the unique hazards and conditions of use at the emergency scene. Also, they must have a large margin of safety because of the likelihood of overloading in an emergency. Industrial and commercial ladders do not require such a margin of safety. Under no circumstances should commercial or household ladders be used for emergency operations.

As with most fire department tools and appliances, frequent concentrated training is required to develop the individual skills and teamwork necessary for efficient use of ground ladders. This is especially true in the case of ground ladders, because the ground ladder is an item of equipment upon which the life of both the firefighter and the public may depend. Even when aerial apparatus is part of the fire fighting attack team, ground ladders will still be needed to gain access to locations inaccessible to apparatus.

The use of ladders goes back to early civilizations. The earliest evidence of their use in North America is not certain, but records indicate that crudely built ladders were being used as early as 1200 a.d. by cliff dwellers in what is now Arizona, Colorado, New Mexico, and Utah. These primitive people built cavelike dwellings that were often three or four stories high. At each level, the rooms were set back into the cliff a few feet from the rooms beneath, thus forming a ledge at each level.

Ladders, rather than steps, were used to reach each level, probably because they could be drawn up when necessary to keep out intruders. These ladders consisted of two poles to which rungs were bound with cords or strips of rawhide. A single beam ladder was also used, which consisted of a pole with notches cut into it to serve as steps. Some rope ladders were also used. They were made of hand-woven vegetable fibers, usually yucca and milkweed.

Early ladders were single-section ladders. Available records do not identify when or where the extension ladder originated. Records do show that in the late 1700s it was the custom to have buckets and a ladder hung at a convenient location in a village for use by whoever could be mustered when a fire occurred.

Hand-drawn hook and ladder trucks were evident by the mid-1800s. They consisted of rackings of single and extension ladders on a specially constructed wagon. Thus, it appears that fire service ladders as we know them today began to emerge at this time.

PURPOSE AND SCOPE

A training manual is more useful when it presents a variety of methods for accomplishing a task, allowing the user to select the one best suited to a particular locality. To achieve this purpose, a selection of methods and techniques for handling, raising, and climbing ground ladders have been recommended by a committee of the International Fire Service Training Association (IFSTA).

Other IFSTA training manuals, such as **Essentials of Fire Fighting**, present only basic information on ground ladders because they are broader in scope and so have limited space. The purpose of this manual is to complement and expand this information by providing more detailed in-depth material.

This manual is intended to prepare firefighters to meet all of the ground ladder objectives contained in NFPA 1001, *Standard for Fire Fighter Professional Qualifications*, Levels I and II. The information ranges from design and construction through carries, raises, climbing, and tactical uses.

The requirements for the design and manufacturer's testing of ground ladders are contained in NFPA 1931, *Standard on Design of and Design Verification Tests for Fire Department Ground Ladders*. These requirements are covered in detail in the first part of this manual. Once a fire department purchases a ground ladder and places it into service, testing and maintaining it becomes the fire department's responsibility. These requirements are contained NFPA 1932, *Standard on Use,* *Maintenance, and Service Testing of Fire Department Ground Ladders*. These requirements are also covered in this manual.

Readers who are familiar with the previous edition of this manual will probably notice that many of the special uses for ground ladders contained in Chapter 5 of the previous edition have been omitted from this edition. These "special" uses included things such as shoring trenchs, making bridges, and using the ladder as a battering ram. During the validation process for this edition, it was brought to our attention that most of these special uses voided the manufacturer's warranty on the ladder. Continued use of ladders in this manner (or continued teaching of these uses) places the user in a potentially serious liability situation. Thus, they have been omitted from this edition.

This manual is limited to information on ground ladders. For information on the operation and use of aerial fire apparatus, see IFSTA's **Fire Department Aerial Apparatus** manual.

Chapter 1

Ladder Types
And Ladder Terms

LEARNING OBJECTIVES

This chapter provides information that will assist the reader in meeting the objectives contained in the Ladders section of NFPA 1001, *Standard for Fire Fighter Professional Qualifications* (1992 edition). The objectives contained in this chapter are as follows:

Fire Fighter I

3-11.1 Identify and describe the use of the following types of ladders:
   (a) Folding/attic
   (b) Roof
   (c) Extension
   (d) Straight/wall
   (e) Aerial devices

Chapter 1
Ladder Types and Ladder Terms

Because of the variety of situations and conditions firefighters encounter, there is no single type of ground ladder that meets all needs. IFSTA recognizes seven different types of ground ladders:

- Single or wall ladders
- Roof ladders
- Folding ladders
- Extension ladders
- Pole or bangor ladders
- Combination or A-Frame ladders
- Pompier ladders

It is important that firefighters be familiar with the various types of ladders. An understanding of ladder configuration in relation to designation is the first step toward understanding ladder usage. This manual contains detailed information on the design, capabilities, and uses of these seven types of ground ladders. This manual does not contain information on the various types of aerial ladders and devices used by the fire service. For information on those types of apparatus, consult IFSTA's **Fire Department Aerial Apparatus** manual.

Any type of specialized equipment has a language of its own that is used in its day-to-day operation. Ladders are no exception to this rule. The following sections of this chapter give firefighters a basic description of the seven types of ground ladders covered in this manual, as well as defining important terms and phrases that are associated with ground ladders.

As with most areas of the fire service, there may be some deviations from the given descriptions and terms based on local terminology or practices.

When possible, this manual tries to account for all known variances. However, if your department has different terms than those covered by this manual, they should, by all means, be incorporated into your training programs.

LADDER TYPES

This section contains brief descriptions and illustrations of the seven basic types of ground ladders recognized by IFSTA.

Single Ladders (Wall Ladders)

Single ladders have only one section and are of a fixed length (Figure 1.1). In some jurisdictions these are called wall ladders; however, for the purpose of this manual the term single ladder will be used. Lengths vary from 6 feet (2 m) to 32 feet (10 m) with the more common lengths ranging from 12 feet (4 m) to 20 feet (6 m). These ladders are mainly used for operations involving one- and two-story buildings.

Roof Ladders

Roof ladders are single ladders that have hooks attached to the tip end. The hooks are nested between the beams when the ladder is used as a single ladder (Figure 1.2). These hooks are swiveled out, as needed, to provide a means of anchoring the ladder when the ladder is being used on a sloped roof (Figures 1.3 a and b). Lengths vary from 12 feet (4 m) to 24 feet (8 m), with 14 and 16 feet (4.3 m and 5 m) being the most common sizes.

Folding Ladders

The folding ladder is a special type of single ladder. It has hinged rungs so that it can be folded into a compact assembly with one beam resting against the other (Figure 1.4). This feature allows

Figure 1.1 A single ladder.

Figure 1.3a A roof ladder with the hooks open.

Figure 1.3b The roof ladder is held in place by extending its hooks over the peak of the roof.

the ladder to be carried in narrow hallways and aisles and to be taken around corners, which is not possible with regular single ladders. When there are low ceilings, as in residential structures, the ladder's compactness makes it much easier to get into attic scuttle openings because it can be inserted while still folded and then opened in place. In fact, in many jurisdictions the folding ladder will be commonly referred to as an "attic ladder." This term is really a misnomer, because small extension or combination ladders may also be used for attic ladders.

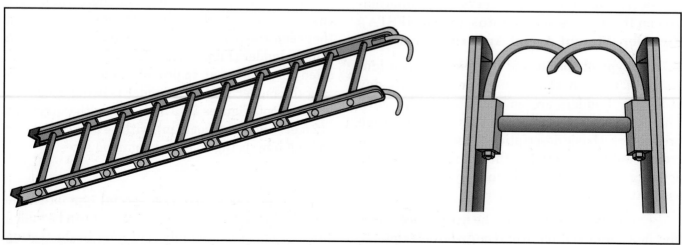

Figure 1.2 A roof ladder with the hooks open and nested.

Folding ladders are narrower when open than regular single ladders (Figure 1.5). This feature allows them to be used in narrow scuttle openings common to residential structures. Lengths range from 8 feet (2.5 m) to 16 feet (5 m) with the most common being 10 feet (3 m). NFPA 1931, *Standard on Design of and Design Verification Tests for Fire Department Ground Ladders*, requires folding ladders to have foot pads attached to the butt to prevent slipping on floor surfaces. The disadvantages of this type ladder are limited weight loading (less than half a regular single ladder) and narrowness, which makes climbing awkward and leg-locking impractical.

Extension Ladders

Extension ladders have either two or three sections. Upper sections are manually raised and lowered by either rope or rope and cable to permit length adjustment. They are referred to by the fully extended length. Lengths range from 12 feet (4 m) to 39 feet (11.5 m) (Figure 1.6).

This ladder's design makes it possible to carry longer ladders on standard-sized apparatus. For example, it would not be practical to carry a 20-foot (6 m) single ladder on the side of a pumper, but a 35-foot (11 m) three-section extension ladder can easily be accommodated.

Pole Ladders (Bangor Ladders)

NFPA 1931 requires permanently attached staypoles on all extension ladders that have a designated length of 40 feet (12 m) or more. These ladders are carried almost exclusively on aerial apparatus and are called pole or bangor ladders, depending on the jurisdiction. The term pole ladder will be used throughout the duration of this manual. The primary purpose for the staypoles is to assist in raising the larger ladders. The staypoles also add stability to the ladder once it has been raised (Figure 1.7).

Pole ladders are manufactured with two to four sections. Lengths vary from 40 feet (12 m) to 65 feet (20 m); however, most modern pole ladders do not exceed 50 feet (15 m). The 55- to 65-foot (17 m to 20 m)

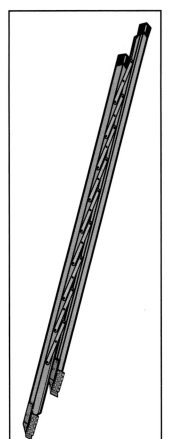

Figure 1.4 In the closed position, a folding ladder is slim and easy to carry.

Figure 1.5 The folding ladder is opened when its position for use is reached.

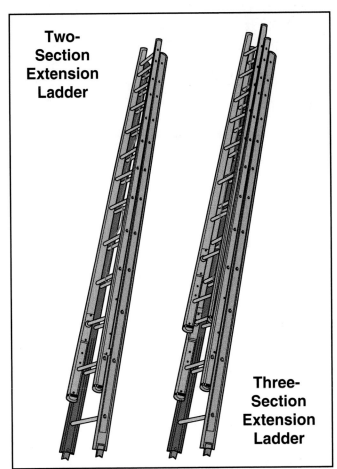

Figure 1.6 Two- and three-section extension ladders.

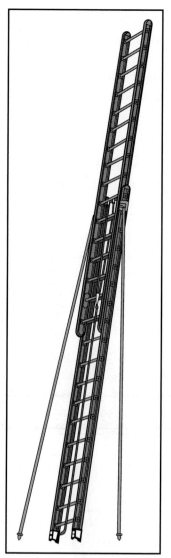

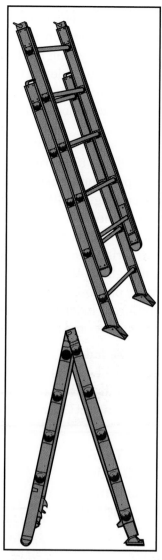

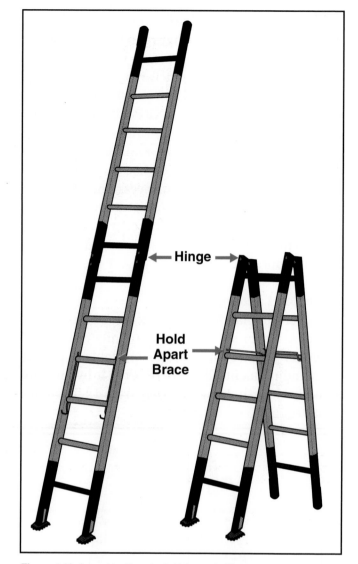

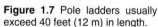

Figure 1.7 Pole ladders usually exceed 40 feet (12 m) in length.

Figure 1.8a A combination extension/A-frame ladder.

Figure 1.8b A combination single/A-frame ladder.

varieties were more commonly found on older aerial apparatus. Pole ladders allow for operations on buildings of up to five stories when aerial apparatus is either not available or unable to gain access to the location where the ladder is required.

Combination Ladders (A-Frame Ladders)

Combination ladders are ladders that can be used both as a self-supported stepladder (A-frame) and as a single or extension ladder (Figures 1.8 a and b). Lengths range from 8 feet (2.5 m) to 14 feet (4.3 m) with the most popular being the 10-foot (3 m) model.

Pompier Ladders

Pompier ladders are single-beam ladders with rungs projecting from both sides. These ladders have a large metal "gooseneck" or hook projecting

from the top of the ladder for insertion into windows or other openings (Figure 1.9). Lengths vary from 10 feet (3 m) to 16 feet (5 m). A few fire departments still use pompier ladders in scaling operations to reach points beyond the range of other ground ladders and aerial apparatus.

LADDER TERMS

Although we usually rely on the Glossary of the manual to provide definitions to key terms, there are some ladder terms that will be used so frequently throughout the remainder of this manual that they warrant coverage in this chapter. The following is a list of commonly used ladder terms.

Angle of Inclination — The preferred pitch for portable, non-self-supporting ground ladders. This preferred pitch is 75.5 degrees (Figure 1.10).

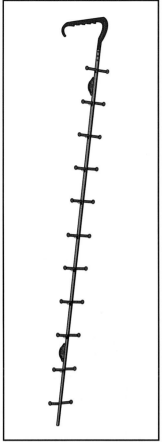

Figure 1.9 Pompier ladders are seldom used in today's fire service.

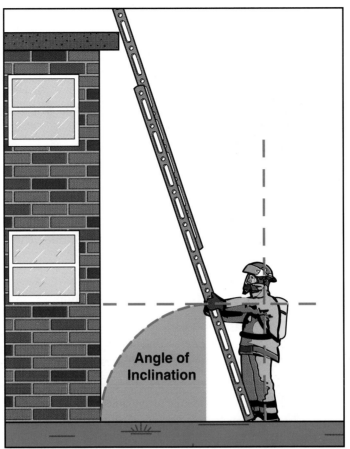

Figure 1.10 The angle of inclination is illustrated here.

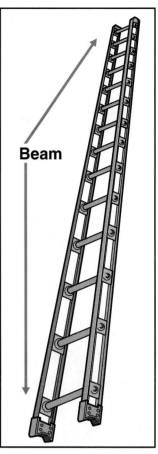

Figure 1.11 A ladder beam.

Beam — Main structural member of a ladder supporting the rungs or rung blocks (Figure 1.11).

Bedded Position — When the fly(s) of an extension ladder is (are) fully retracted it is said to be in the bedded position; the position in which the ladder is carried on the apparatus (Figure 1.12).

Bed Section (Base Section) — The lowest or widest section of an extension ladder (Figure 1.13). This section always

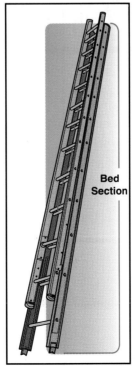

Figure 1.13 The bed section is the widest section of an extension ladder.

maintains contact with the ground or other supporting surface.

Butt — The bottom end of the ladder; the end which will be placed on the ground or other supporting surface when the ladder is raised. Also called Heel (Figure 1.14).

Butt Spur — Metal safety plates or spikes attached to the butt of ground ladder beams to prevent slippage (Figure 1.15).

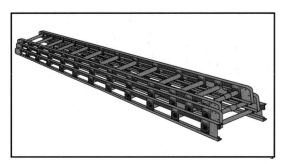

Figure 1.12 An extension ladder in the bedded position.

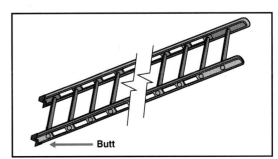

Figure 1.14 The bottom of the ladder is known as the butt.

Figure 1.15 The butt spur helps keep the ladder anchored on soft surfaces.

Figure 1.16 The length label is affixed to the outside of the ladder within 12 inches (300 mm) of the butt.

Designated Length — The length marked on the ladder; markings shall be within 12 inches (300 mm) of the butt of each side rail of a single ladder and the base section of extension and pole ladders. These markings shall be visible when the ladder is in the bedded position and stored on the apparatus (Figure 1.16).

Dogs — See Pawls.

Fly Section — Upper section(s) of extension, pole, or some combination ladders (Figure 1.17).

Guides — Wood or metal strips, sometimes in the form of slots or channels, on an extension ladder that guide the fly section while being raised.

Halyard — Rope used on extension ladders to extend the fly sections (Figure 1.18). Also called Fly Rope.

Heel — See Butt.

Identification Number — The serial number placed on each ground ladder by the manufacturer as required by NFPA.

Inside Width — The distance measured from the inside of one beam to the inside of the opposite beam (Figure 1.19).

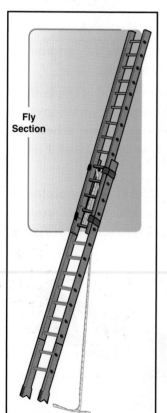

Fly Section

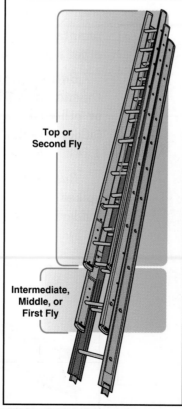

Top or Second Fly

Intermediate, Middle, or First Fly

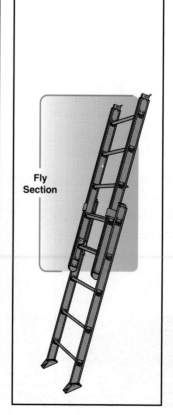

Fly Section

Figure 1.17 Fly sections of extension and combination ladders.

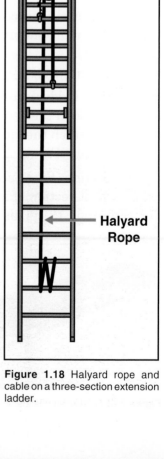

Halyard Cable

Halyard Rope

Figure 1.18 Halyard rope and cable on a three-section extension ladder.

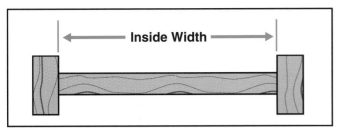

Figure 1.19 The inside width of a ladder.

Maximum Extended Length — The total length of the extension, pole, or some combination ladders when all fly sections are fully extended and the pawls are engaged (Figure 1.20).

Nesting — The procedures whereby ladders of different sizes and/or types are racked partially within one another to reduce the space required for storage on the apparatus. The most common arrangement is to nest the roof ladder with an extension ladder on the side of a pumper (Figure 1.21).

Figure 1.20 This figure illustrates the maximum extended length of an extension ladder.

Figure 1.21 A common nesting arrangement for ladders.

Outside Width — The distance measured from the outside of one ladder beam to the outside of the opposite ladder beam (Figure 1.22).

Pawls — Devices attached to the inside of the beams on fly sections used to hold the fly section in place after it has been extended (Figure 1.23). Also called Dogs or Ladder Locks.

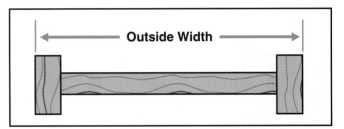

Figure 1.22 The outside width of a ladder.

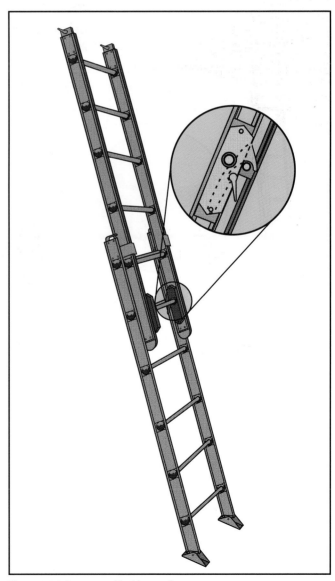

Figure 1.23 One type of ladder pawls.

Rail — The side rail of a ladder. Also called Beam.

Retracted — See Bedded Position.

Rungs — Cross members which provide the foothold for climbing. In all except pompier ladders, the rungs extend from one beam to the other; on a pompier ladder the rungs pierce the single beam (Figure 1.24).

Side Rail — The side rail of a ladder. Also called Rail or Beam.

Tip (Top) — The extreme top of the ladder (Figure 1.25).

It is important for firefighters to have a thorough knowledge of all of these ladder terms. Knowledge of these terms will speed the selection, use, cleaning, maintenance, and inspection of ladders.

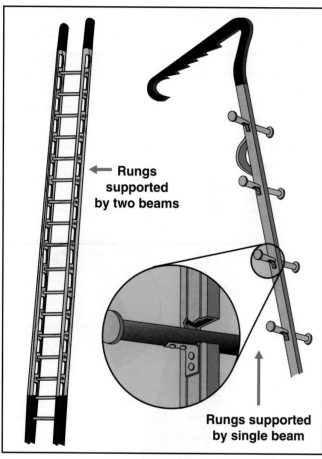

Rungs supported by two beams

Rungs supported by single beam

Figure 1.24 Regular and pompier ladder rungs.

Tip

Figure 1.25 The top of the ladder is known as the tip.

Chapter 1 Review

Directions

The following activities are designed to help you comprehend and apply the information in Chapter 1 of **Fire Service Ground Ladders,** Ninth Edition. To receive the maximum learning experience from these activities, it is recommended that you use the following procedure:

1. Read the chapter, underlining or highlighting important terms, topics, and subject matter. Study the photographs and illustrations, and read the captions under each.

2. Review the list of vocabulary words to ensure that you know the chapter-related meaning of each. If you are unsure of the meaning of a vocabulary word, look the word up in the glossary or a dictionary, and then study its context in the chapter.

3. On a separate sheet of paper, complete all assigned or selected application and review activities before checking your answers.

4. After you have finished, check your answers against those on the pages referenced in parentheses.

5. Correct any incorrect answers, and review material that was answered incorrectly.

Vocabulary

Be sure that you know the chapter-related meanings of the following words.

- retracted *(9)*
- bedded *(9)*
- nested *(5)*
- scuttle *(6)*
- pitch *(8)*

Application Of Knowledge

1. Make a chart in which you provide the following information for each of the seven ground ladders recognized by IFSTA: configuration and design, primary uses, lengths, and miscellaneous.

2. Draw or trace a simple picture of an extension ladder, and then label its parts.

3. Examine examples of ground ladders used in your fire department.

Review Activities

1. List the seven different types of ground ladders recognized by IFSTA. *(5)*

2. Briefly describe the configuration and design of each of the ground ladders listed in Activity 1. *(5-8)*

3. Identify the primary uses for each of the ground ladders listed in Activity 1. *(5-8)*

4. Identify the following ladder parts:
   - beam *(9)*
   - bed section *(9)*
   - butt *(9)*
   - butt spur *(9)*
   - dogs *(10)*
   - fly section *(10)*
   - foot pad *(7)*
   - gooseneck *(8)*
   - guides *(10)*
   - halyard *(10)*
   - heel *(10)*
   - pawl *(11)*
   - rail *(12)*
   - rungs *(12)*
   - side rail *(12)*
   - staypole *(7)*
   - tip *(5)*

5. Identify the following terms associated with ground ladders:
   - identification number *(10)*
   - inside width *(10)*
   - designated length *(10)*
   - maximum extended length *(11)*
   - outside width *(11)*
   - angle of inclination *(8)*

GROUND LADDERS

Chapter 2

Ground Ladder Construction, Maintenance, and Service Testing

LEARNING OBJECTIVES

This chapter provides information that will assist the reader in meeting the objectives contained in the Ladders section of NFPA 1001, *Standard for Fire Fighter Professional Qualifications* (1992 edition). The objectives contained in this chapter are as follows:

Fire Fighter II

4-11.1 Identify the materials used in ladder construction.

4-11.2 Identify the load capacities established by NFPA 1931, *Standard on Design of and Design Verification Tests for Fire Department Ground Ladders,* and NFPA 1904, *Standard for Aerial Ladder and Elevating Platform Fire Apparatus,* for ground and aerial ladders.

4-11.3 Demonstrate the procedures for cleaning ladders.

4-11.4 Demonstrate inspection and maintenance procedures for different types of ground and aerial ladders.

4-11.5 Describe the annual service test for ground ladders.

Chapter 2
Ground Ladder Construction, Maintenance, and Service Testing

You will almost always find a wide variance in the design and materials used in the construction of any piece of fire service equipment. Ground ladders are certainly no exception to this rule. When selecting ground ladders, fire service purchasing agents have several choices for both the construction materials and the design of the ladder. This chapter will highlight all the major construction methods of ground ladders in use by today's fire service.

Regardless of what material or design is used for the ladder, it must conform to minimum performance standards established by the National Fire Protection Association (NFPA). NFPA 1931, *Standard on Design of and Design Verification Tests for Fire Department Ground Ladders,* lays out the basic design requirements for all ground ladders, regardless of their type of construction. Table 2.1 contains the load capacity requirements for each major type of ladder.

TABLE 2.1 Maximum Ladder Loading According To NFPA 1931 (1994 Edition)		
Type of Ladder	**Load**	
	US	**Metric**
Folding	300 pounds	(136 kg)
Pompier	300 pounds	(136 kg)
Single and Roof	750 pounds	(340 kg)
Extension and Pole	750 pounds	(340 kg)
Combination	750 pounds	(340 kg)

Fire department ground ladders, like any tool or appliance, require unique maintenance procedures. These will vary according to the type of construction and materials used. The importance of maintenance is critical in the case of ground ladders because of their direct involvement in life-saving situations. Most of this maintenance is performed by the firefighter who is, after all, one of the main beneficiaries of the effort involved. The maintenance section of this chapter deals with this aspect.

Ground ladders, like all other fire fighting tools and apparatus, are required to be service tested at specified intervals. The procedures for service testing ground ladders will be examined later in this chapter. However, before discussing service testing, firefighters need to be familiar with the construction of these ladders. NFPA 1932, *Standard on Use, Maintenance, and Service Testing of Fire Department Ground Ladders,* provides the requirements for both maintaining and service testing ground ladders.

SINGLE, ROOF, EXTENSION, AND POLE LADDER CONSTRUCTION

Single, roof, extension, and pole ladders are the most commonly used fire service ground ladders. Similar materials and design methods are used for all of these ladders. The following sections examine these materials and designs.

Fire service ground ladders have to be able to withstand considerable abuse, such as:

- Overloading (Figure 2.1)
- Temperature extremes
- Being engulfed in flames
- Structural collapse

Figure 2.1 Ladders can be damaged by having too many people on them at the same time.

Because ladders are readily available at the fire scene, they are sometimes used for tasks for which they were not designed. When this is the case, firefighters must be careful not to exceed the ladder manufacturers' recommendations for safe use (Figure 2.2).

Figure 2.2 Another way to damage a ladder is to use it for a purpose other than what it was designed for, such as shoring a trench.

Every time a fire service ladder is used, someone's life (most frequently that of the firefighter) depends on it functioning properly. Because of this, it is very important that the ladder have no structural defects or design weaknesses. Workmanship should also be such that other defects that may cut or tear clothing or skin do not exist. It was for these reasons that NFPA 1931 was developed and adopted. Fire service ground ladders should be purchased from specifications that adopt NFPA 1931 by reference.

Firefighters and chief officers can be assured that a ladder is safe for fireground use if it meets the requirements of NFPA 1931. Ladders that meet this standard can be easily identified by the presence of a label affixed directly to the ladder (Figure 2.3). Information on what tests the ladder must pass to carry this label are discussed briefly later in this chapter.

Figure 2.3 The manufacturer should affix a label to the ladder that certifies the ladder was constructed in accordance with NFPA 1931.

There are three basic materials used to construct fire service ground ladders: Wood, metal, and fiberglass. The following sections describe each of these in detail.

Wood Ladders

Wood was the first material used for the construction of ground ladders (Figure 2.4). Although metal and fiberglass ladders have become very popular with the fire service, many fire departments still choose wood ladders.

Figure 2.4 Some fire departments still choose to use wood ground ladders.

The wood used for fire service ladders must be carefully selected before the manufacturing process begins. Only a few types of lumber meet the requirements of weight per cubic inch, bending strength, stiffness, hardness, and resistance to shock. The most desirable of these are clear straight-grained Coast Douglas fir for beam construction, hickory for rungs, and red oak for certain critical wear parts and stops.

Coast Douglas fir is used because strength, toughness, and flexibility are combined with relatively light weight. The use of a select grade eliminates knots, checks, pitch pockets, and other naturally occurring undesirable characteristics. The moisture content of the wood must be reduced to 12 percent by air drying, kiln drying, or a combination of both before the wood is used for ladder construction.

Not just any Douglas fir is used; it must be Coast Douglas fir, which grows in climate conditions that result in growth rings relatively close together. Wood that is sawed so that the grain is angular provides maximum strength in all directions (Figure 2.5). Such a piece is described as having vertical grain on all four sides. This piece would be used for the side rails of a ground ladder.

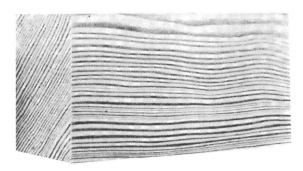

Figure 2.5 Ladder side rail stock sawed so that the grain is angular. Sawing in this manner provides maximum strength.

Wood that is sawed with the grain flat has more strength in one direction than the other (Figure 2.6). It has flat gain on the top and bottom surfaces and vertical grain on the side surfaces. A sectional view of a truss beam for a wood ladder should show each piece having vertical grain on all four sides (Figure 2.7).

An enlargement of a piece of Coast Douglas fir is shown in Figure 2.8. Note the cellular structure of the wood. Although all wood is not the same, the enlargement shows that wood is porous like a honeycomb. The dark diagonal streaks, which are the dense summer wood, are generally termed the growth rings. It is not difficult to understand how excessive loss of moisture, resins, and oils can leave empty cells that result in shrinkage and loss of strength.

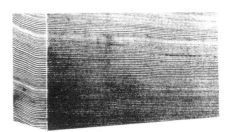

Figure 2.6 Ladder side rail stock sawed with the grain flat. The wood is not as strong when sawed this way and does not wear well.

Figure 2.7 A sectional view of a wood truss beam.

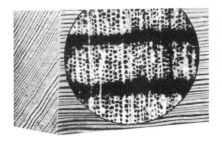

Figure 2.8 An enlargement of a piece of Douglas fir. Note the cellular structure and the porous honeycomblike composition.

Wood shrinkage is particularly noticeable after a wood ladder has been subjected to both low humidity and artificial heat — conditions that are common in fire stations. This slow drying process is a primary cause of loose rungs and cracked rails in wood ladders.

There are two primary types of construction of wood ground ladders: solid beam and truss beam. The following sections go into a little more detail on beam and rung materials and construction.

SOLID BEAM CONSTRUCTION

The solid beam of a wood ladder is just what the term implies: a single piece of solid wood, usually Coast Douglas fir. Solid beam extension ladders use a metal guide attached to the beam of the fly section near the bottom and another guide attached to the bed section near the top to hold the sections together (Figure 2.9). The rungs are typically attached to the center of the solid beam.

Figure 2.9 Note the guides attached to both sections of this solid beam wood extension ladder.

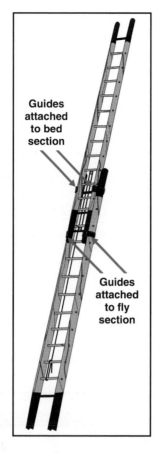

Guides attached to bed section

Guides attached to fly section

TRUSS BEAM CONSTRUCTION

Truss beam ladders have two rails separated by truss blocks. These ladders are usually constructed of Coast Douglas fir. For wood ladders exceeding 24 feet (8 m) in length, truss beam construction will produce a much lighter ladder than will solid beam construction. This is the primary advantage of truss beam construction over solid beam construction. There are two basic truss beam designs used for fire service ground ladders:

- Those with rungs mounted in the truss blocks

- Those with rungs mounted in the top rail

Rungs Mounted In Truss Blocks

There are two styles of ladders that have rungs mounted in the truss blocks: those with one piece rung and truss block construction and those with separate truss blocks (Figure 2.10). There are three variations of those with separate truss blocks:

- Rung centered in the truss block. Identical size parallel rails (Figure 2.11).

- Rung mounted at bottom center of truss block adjacent to the bottom truss rail. Identical size parallel rails (Figure 2.12).

- Rungs mounted at bottom center of truss block adjacent to the bottom truss rail. Identical size rails. Top rail arches. Bottom rail is straight (Figure 2.13).

Extension ladders with rungs mounted in the truss blocks are designed to be used with the fly out. More information on fly position can be found in Chapter 4 of this manual.

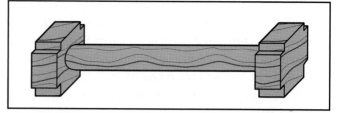

Figure 2.10 A typical one-piece rung and truss block construction.

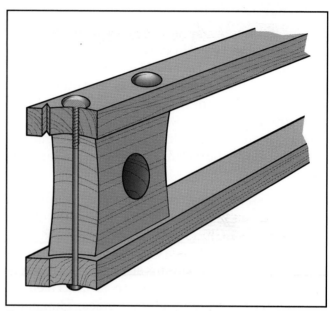

Figure 2.11 This design incorporates rungs mounted in the center of the truss block.

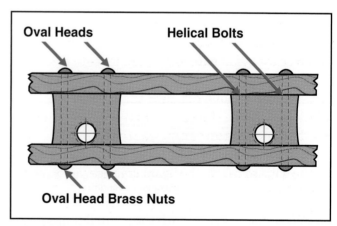

Oval Heads   Helical Bolts

Oval Head Brass Nuts

Figure 2.12 Identical parallel beam rails with rungs mounted in the center bottom of the truss blocks.

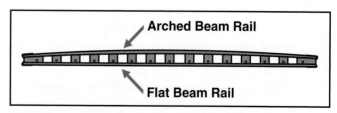

Arched Beam Rail

Flat Beam Rail

Figure 2.13 A ladder where one beam rail is flat and the other is arched; the rungs are mounted at bottom center of the truss block adjacent to the bottom truss rail.

Rungs Mounted In Top Rail

This design has a thicker top rail and, except for the roof ladder, both rails arch to form tapered trusses (Figure 2.14). These types of ladders are designed to be used with the fly in. The bottom rail of the roof ladder is straight so that it will lie flat on a roof (Figure 2.15).

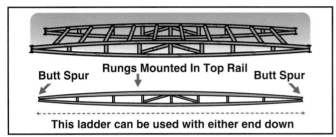

Figure 2.14 This design uses a thicker top rail to support the rungs. Both beam rails arch on wall ladders of this type, and the rungs are mounted in the top rail.

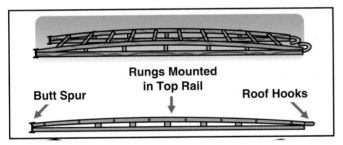

Figure 2.15 Although the rungs are mounted in the top rail of this ladder, the bottom rail is flat so that it will lie on the roof properly.

One method used to hold sections together on wood truss beam extension ladders is to construct the fly beam with a wooden tongue that travels in a U-shaped groove cut into the side of the truss block just above the rungs (Figure 2.16). Another method employs an oak strip attached to the inside of the beam (Figure 2.17). When rungs are mounted in the bottom beam, the beam rails are made wider than the truss blocks. The resulting groove acts as the guide and holds the assembly together (Figure 2.18).

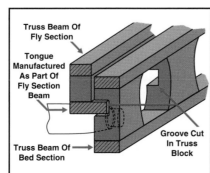

Figure 2.16 A tongue manufactured as part of the bottom truss rail of a fly section travels in a groove cut into the truss blocks of the bed section rail.

Figure 2.17 An oak strip is attached to the inside of the bed section beam.

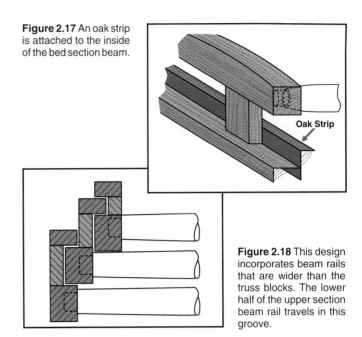

Oak Strip

Figure 2.18 This design incorporates beam rails that are wider than the truss blocks. The lower half of the upper section beam rail travels in this groove.

RUNGS

To meet requirements of NFPA 1931, rungs for both solid and truss beam ladders must be a minimum of 1¼ inches (32 mm) in diameter at the center. Most are the same diameter all the way across, but some are thicker in the center. These are called swell center rungs (Figure 2.19). Swell center rungs must be 1¼ inches (32 mm) in diameter at the center but can taper to 1⅛ inches (29 mm) at the ends.

NFPA 1931 calls for rungs to be spaced between 12 and 14 inches (305 mm and 356 mm) on center, plus or minus ⅛-inch (3 mm). No tread surface is required for wooden rungs.

There are three common rung installation methods. In all three methods, the rung is manufactured with a tenon on each end. The first method uses a tenon set in a mortise drilled in the ladder beam rail

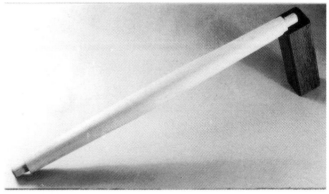

Figure 2.19 A swell center rung. *Courtesy of ALACO Ladder Company.*

(Figure 2.20). When this method is used, the ladder tie rods keep the beams tight against the rung shoulder. The rung is prevented from turning by a metal key driven into the rung tenon when the tenon is inserted into the ladder beam rail. A nail or wood screw may also be used for this purpose.

The second method has the full diameter of the rung entering the beam rail (Figure 2.21). It is said to be stronger than the first method described. Tie rods keep the beams bearing on the rung, and a key or screw keeps the rung from turning. However, the taper of the rung will usually be sufficient to prevent the rung from turning. Some rungs are held in place by gluing (Figures 2.22 a and b). The swell center rung pictured in Figure 2.19 is installed in this manner.

Some rungs are both glued and nailed. Rungs set in truss blocks are attached in much the same method except that the tenon passes through the truss block. Its end is flush with the outside surface of the truss block. When the rung and truss block are one piece, the truss block is keyed to both the upper and lower beam rails, and a single beam bolt is used to hold it in place (Figure 2.23).

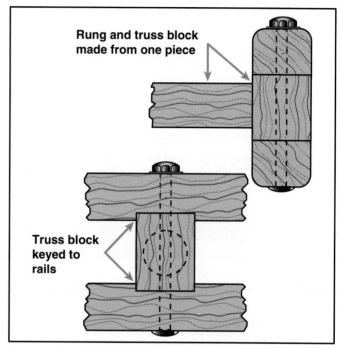

Figure 2.23 When the rung and truss block are one piece, the upper and lower beam rails are keyed and a single beam bolt is used.

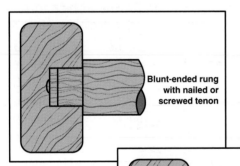

Figure 2.20 A blunt tenon-in-mortise mounting arrangement.

Blunt-ended rung with nailed or screwed tenon

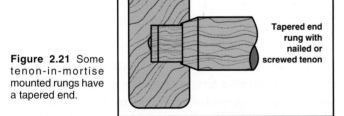

Figure 2.21 Some tenon-in-mortise mounted rungs have a tapered end.

Tapered end rung with nailed or screwed tenon

Figure 2.22a One style of glued tenon rung mount. *Courtesy of ALACO Ladder Company.*

Figure 2.22b Another style of glued tenon rung mount. *Courtesy of ALACO Ladder Company.*

ADVANTAGES OF WOOD LADDERS

Advantages of wood ground ladders include the following:

- When dry and clean, wood ladders are non-conductors of electricity. This makes them much safer than metal ladders for use around live electrical power sources.

- Wood ladders are not prone to sudden failure when subjected to heat or flames.

- Wood is a good heat insulator; it does not transmit temperature extremes like metal does.

DISADVANTAGES OF WOOD LADDERS

Disadvantages of wood ground ladders include the following:

- A finish, such as a spar varnish, is required to protect the wood. This finish must be kept intact, which often requires considerable time and labor.

- Wood ladders deteriorate with age.

- Wet or dirty ladders will conduct electricity.

- Usually, wood ladders are heavier than metal and fiberglass ladders. However, this may not always be the case.

Metal Ladders

Until WWII, nearly all fire service ladders were constructed of wood. Metal ladders were then introduced. They became popular because they were cheaper and lighter in weight. Today, the metal ladder is the most widely used by the fire service even though current standards necessitate their being heavier than in the past (Figure 2.24).

Figure 2.24 Metal ladders are the most common type used in today's fire service. *Courtesy of Joel Woods.*

Aluminum is the metal of choice used for the construction of ground ladders. Usually, aluminum is the primary ingredient in an alloy of several metals that are used to construct the ladder. Because of this, many fire service personnel commonly refer to these ladders as "aluminum ladders." However, because the alloys vary widely, IFSTA prefers the term "metal ladder" for all ladders of metallic construction.

NFPA 1931 requires that all structural components of metal ladders be manufactured of materials that maintain at least 75 percent of their designated design strength at a minimum of 300°F (149°C). As with wood ladders, metal ladders are available in both solid and truss beam construction.

SOLID BEAM CONSTRUCTION

The term "solid beam" when used to refer to a metal ladder is actually a misnomer because metal

"solid beams" are not solid in the sense that wood beams are. One design uses a C-channel. Rung plates are riveted across the open side of the "C" to provide support and to serve as mounting for the rungs (Figure 2.25). Some extension ladders made with this type beam construction have modified rung plates for bed and intermediate fly sections (Figure 2.26). The beam rail of the fly section is modified by adding a "tongue" (Figure 2.27). The resulting assembly is called tongue and groove construction (Figure 2.28). Others use the same guide design that was originally developed for wood ladders. A guide is riveted to the fly section beam near the bottom, and another guide is riveted to the bed section beam near the top (Figure 2.29).

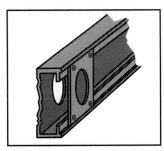

Figure 2.25 C-channel beam rail construction. Rung plates are riveted across the open side of the C.

Figure 2.26 Rung plate modification for the C-channel beam rail of the bed section of an extension ladder.

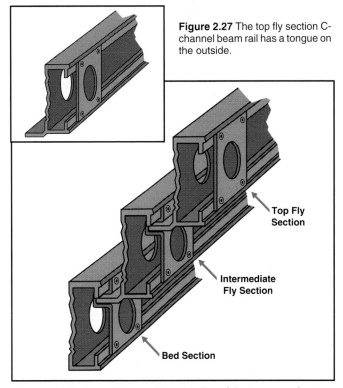

Figure 2.27 The top fly section C-channel beam rail has a tongue on the outside.

Top Fly Section

Intermediate Fly Section

Bed Section

Figure 2.28 C-channel beam rail construction of the tongue and groove type with three sections in place.

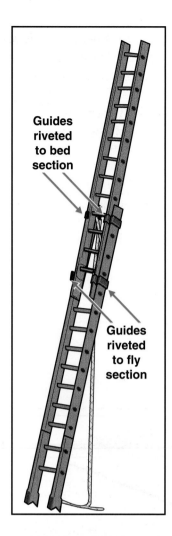

Guides
riveted
to bed
section

Guides
riveted
to fly
section

Figure 2.29 The guides are usually riveted to the beams.

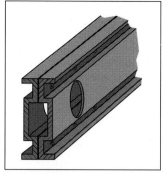

Figure 2.30 Two extruded structural members are riveted together to form this beam rail.

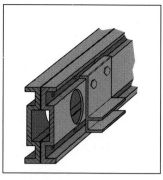

Figure 2.31 U-channel guides bolted to the outside of the fly section near its beam.

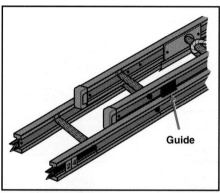

Guide

Figure 2.32 Bed and fly section assembled to form an extension ladder.

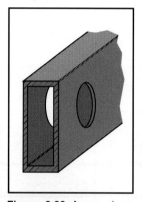

Figure 2.33 A one-piece tubular beam rail for a single or roof ladder.

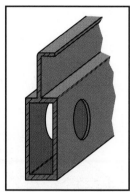

Figure 2.34 A one-piece extruded tubular beam for the bed section of an extension ladder. Protruding ends of the rungs of the fly section travel in the channel.

Another design uses a tubular beam rail made by riveting two extruded structural members together (Figure 2.30). Extension ladders with this type beam rail construction have a plate with a U-channel bolted to the outside of the fly beam near the bottom of the fly section (Figure 2.31). A similar plate and U-channel are bolted to the inside of the bed ladder beam near the top. The top T-rail of the bed section slides through the U-channel of the fly section. The bottom T-rail of the fly section beam rail slides through the U-channel attached to the top of the bed ladder beam rail (Figure 2.32). A one-piece, extruded tubular beam is also available (Figure 2.33).

Extension ladders are manufactured with a C-channel as a part of the top of the bed section beam rail (Figure 2.34). Rungs of the fly section protrude through the outside of the beam rail and slide up and down in the C-channel. The protruding portions of the rungs are fitted with nylon bushings to make them slide easier (Figure 2.35). The completed assembly is shown in Figure 2.36.

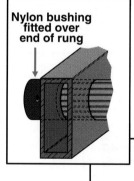

Nylon bushing
fitted over
end of rung

Figure 2.35 The protruding ends of the fly section rungs are fitted with nylon bushings to make them slide easier.

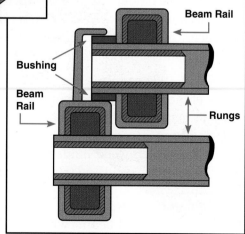

Beam Rail

Bushing

Beam
Rail

Rungs

Figure 2.36 A metal extension ladder assembly.

TRUSS BEAM CONSTRUCTION

Truss beam construction also varies. One type uses two rectangular rails for each beam. The rails have a U-shaped channel manufactured on the facing surfaces of each rail. Truss plates are riveted between the U-channels to hold the two rails together and to form a beam (Figure 2.37).

Extension ladders with this type beam construction are made so that the outside width of the fly is slightly less than the inside width of the bed section. U-channels are attached to the side of the bed section truss plates (Figure 2.38). These channels act as a guide for the rail of the fly section to slide in as it is extended and serve to hold the assembly together. When there is another fly, as in a 35-foot (11 m) three-piece extension ladder, the scheme is repeated (Figure 2.39).

The second type of metal truss beam construction uses two T-channels joined by riveted truss plates (Figure 2.40). The rungs of this extension ladder protrude from the outside of the beams of the fly section(s) and slide in a U-shaped guide manufactured as a part of the side of the upper T-rail (Figure 2.41).

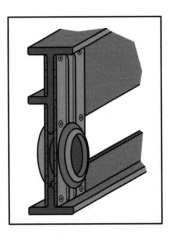

Figure 2.40 Double T-channel truss beam construction. The bed section of an extension ladder is shown. The ends of rungs protruding from the fly section rail fit into the U-channel on the inside of the top rail.

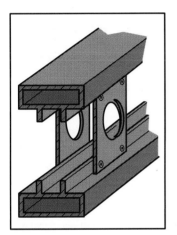

Figure 2.37 Double rectangular rail truss beam construction.

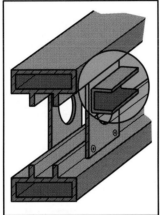

Figure 2.38 The U-shaped guide on the bed section of an extension ladder.

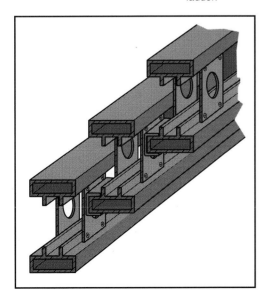

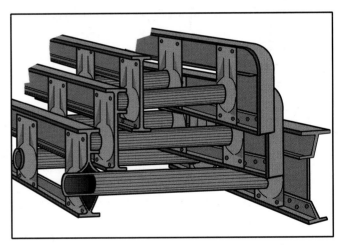

Figure 2.41 Fly section rungs protrude beyond the outside of the beam rail and ride in U-shaped channels that are part of the top T-rail.

RUNGS

Metal rungs are used on metal ladders. The requirements in NFPA 1931 for metal rungs are essentially the same as those described for wood ladders. They must be spaced on 14-inch (356 mm) centers plus or minus ⅛-inch (3 mm). Metal rungs must be constructed of heavy-duty corrugated, serrated, knurled, or dimpled material, or they must be coated with a skid-resistant material.

A variety of methods are used to attach rungs to beams; most use the truss or rung plate as the point of support. In the case of the C-channel "solid" beam construction, a rung plate is riveted across the open side of the C. Holes the size of the rungs are cut through the rung plate and the ladder beam, and the rung is inserted. It is then expanded on both sides of the rung plate, and the end is welded flush to the outside of the beam (Figure 2.42).

A second design of similar nature is used for truss ladder construction except that the rung end is welded to the outside of the outer truss plate (Figure 2.43). A third design uses a bushing that is inserted into the end of the rung and then expanded into place (Figure 2.44).

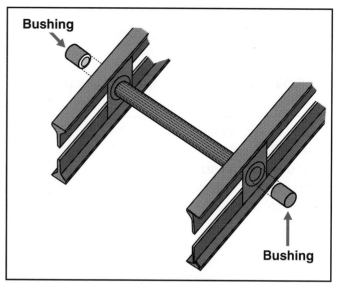

Figure 2.44 This design secures the rung by using bushings that are inserted into the rung ends and then expanded.

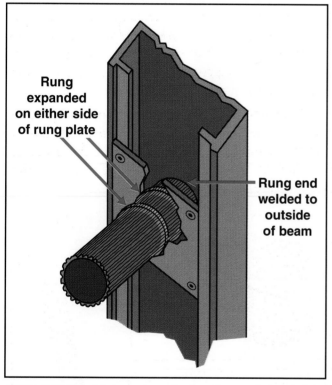

Figure 2.42 With this type of construction, the rung is inserted and expanded on either side of the rung plate. The end of the rung is welded to the outside of the beam rail.

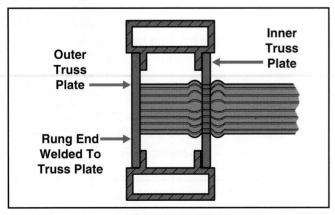

Figure 2.43 This design is used for truss ladders. The rung is expanded on both sides of the inner truss plate, and it is welded to the outer truss plate.

ADVANTAGES OF METAL LADDERS

Other than the fact that metal ladders are more readily available than wood ladders, metal ladders have been found to have some definite advantages:

- In most lengths and models, metal ladders are lighter in weight than their wood and fiberglass counterparts. However, there are a few exceptions. Table 2.2 uses a 24-foot (8 m) extension ladder as an example of the weight differences between ladders of various construction types. This information is obtained from ladder manufacturers' catalogs. Weight differences for other types and sizes of ladders are proportionate to these figures.

- Metal ladders are tougher than either wood or fiberglass, so they show less wear and tear from everyday use.

- They are not as susceptible to chipping and cracking when subjected to impact as are wood and fiberglass ladders.

- Users are not subject to injury from splinters, as is sometimes the case with wood ladders.

- Metal ladders are easier to care for than wood ladders because they do not need to be sanded down and refinished periodically. Since this refinishing is a time-consuming process, metal ladders have less downtime.

- Metal ladders do not deteriorate with age and are not subject to absorption and dry rot.

TABLE 2.2 24-Foot (8 m) Ladder Weights		
Type of Ladder	**Weight**	
	(lb)	**(kg)**
Solid Beam, Metal	74	34
Truss Beam, Metal	97	44
Solid Beam, Wood	75	34.5
Truss Beam, Wood	110	50
Fiberglass	85	39

DISADVANTAGES OF METAL LADDERS

There are also some disadvantages to using metal ladders:

- First and foremost, they are good conductors of electricity, even when dry.

WARNING

Extreme caution is necessary whenever metal ladders are used near electrical power sources. Contact with power sources may result in electrocution of anyone in contact with the ladder.

- Metal ladders are subject to sudden failure when exposed to heat or flame temperatures of 300°F (149°C) or more, even for short periods. These kinds of temperatures are routinely encountered at fires. Even though they may not fail immediately, the metal loses its strength when so heated and does not regain it when cool.

WARNING

Any metal ladder subjected to direct flame contact or heat high enough to cause water contacting it to sizzle or turn to steam or whose heat sensor label has changed color should be removed from service and tested.

- Metal ladders may become very cold in winter or very hot in summer because of the good conductivity of the aluminum alloy.

Fiberglass Ladders

Fiberglass is one of the newer materials used for construction of fire service ground ladders. Its acceptance has been slow because it is heavier than corresponding metal and wood ladders. Fiberglass ladders are not actually all fiberglass; the beams are fiberglass and the remaining parts are metal.

Fiberglass fire service ladders are manufactured only in solid beam construction. The beam is not actually solid, but like some metal ladders it employs a C-channel construction. A metal rung plate is riveted across the open side of the C-channel, and the rung, which is metal, is attached to this plate just as it is in metal construction. The rungs are constructed the same as for metal ladders, and the spacing requirement is the same (Figure 2.45).

Extension ladders use guides identical to solid beam wood ladders. One guide is attached to the fly section beam near the bottom, and the other is attached to the bed section beam near the top (Figure 2.46).

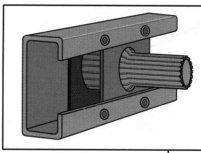

Figure 2.45 Rung attachment for a fiberglass ladder.

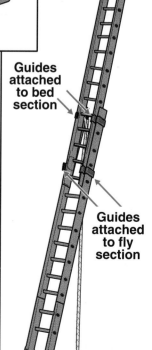

Guides attached to bed section

Guides attached to fly section

Figure 2.46 The guides used on fiberglass ladders are similar to those used on metal and wood ladders.

ADVANTAGES OF FIBERGLASS LADDERS

The following properties are advantages of fiberglass ladders:

- Fiberglass is a nonconductor of electricity when dry.

- Fiberglass can be exposed to low heat levels without losing its original load-carrying capacity.

- Fiberglass is a tough material that can take considerable abuse but requires little maintenance.

- Fiberglass is a poor conductor of heat.

DISADVANTAGES OF FIBERGLASS LADDERS

The disadvantages of fiberglass construction include the following:

- Fiberglass is a dense material and so is relatively heavy.

- Fiberglass tends to chip and crack with severe impact.

Composite Wood and Metal Ladders

Ladders constructed of a combination of wood beams and metal rungs have also been marketed, primarily to meet the objections that some have to the electrical conductivity hazard of metal ladders.

Composite ladders are manufactured only in truss beam construction. The rails are made of Coast Douglas fir. Metal truss plates are used to complete the beam construction. The rails are held together by two beam bolts extending through the beam at each truss plate. Metal rungs of the standard size and spacing are used and are welded to the outer side of the truss plate. Ribs are expanded on either side of the inner truss plate. On extension ladders, the sections are held together by interlocking the bottom rail of the fly with the top rail of the bed ladder (Figure 2.47).

ADVANTAGES OF COMPOSITE WOOD AND METAL LADDERS

There are two primary advantages of composite wood and metal ladders: First, the wood beams are a nonconductor of electricity when clean and dry. Secondly, they are slightly lighter than comparable all-wood ladders.

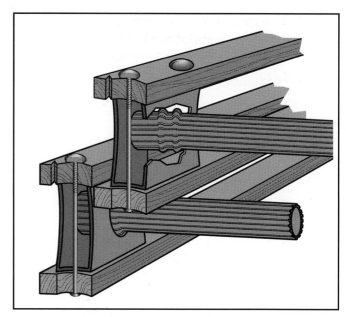

Figure 2.47 Composite ladders are held together by interlocking the upper truss beam rail of the bed section with the lower truss beam rail of the upper section.

DISADVANTAGES OF COMPOSITE WOOD AND METAL LADDERS

The primary disadvantage of composite wood and metal ladders is that the beams have all the wear and maintenance disadvantages of a wood ladder. They may require considerable maintenance to keep them in service.

OTHER TYPES OF LADDER CONSTRUCTION

In addition to the ladders described earlier, a variety of other types of ladders are used by the fire service. These include folding, combination, and pompier ladders. The following sections highlight the major characteristics of these ladders.

Folding Ladders

Folding ladders may be constructed of metal, wood, or fiberglass. Of the three, metal construction is most common. Folding ladders range in length from 6 to 14 feet (2 m to 4.3 m). There are two primary construction differences between these and other fire service ground ladders:

- They are required to have foot pads to prevent slippage.

- The rungs are hinged at both ends so that the ladder can be folded into a compact assembly.

Beams of metal folding ladders are manufactured in three designs. One is a tubular rail with a

U-channel manufactured on one side. Each rung consists of two flat metal strips held in place against each side of the U-channel by a single hinge pin (Figure 2.48).

A second design uses a U-channel beam with a smaller U-channel in the center to support the rungs. Rungs are of square tubular construction and are attached to the smaller U-channel by a hinge pin (Figure 2.49).

A third design incorporates a C-channel beam with square tubular rungs attached to the C-channel by hinge pins (Figure 2.50). All three designs use a bracelike device to lock the ladder in the open position (Figure 2.51).

Wood folding ladders use two parallel wood beams separated only by the width of the wood rung; no truss block is used. The rung hinge pin holds the assembly together. The rungs are square (Figure 2.52).

Combination Ladders

Most combination ladders are of metal construction, but some are made of wood or fiberglass. Beam and rung construction, unless specifically

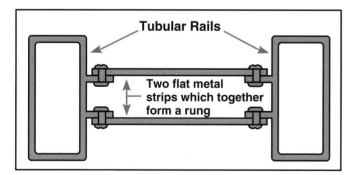

Figure 2.48 This folding ladder is constructed of tubular rails with parallel metal strips for rungs.

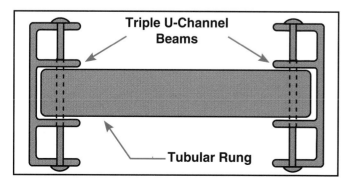

Figure 2.49 This design uses a triple U-channel beam with rectangular tubular rungs.

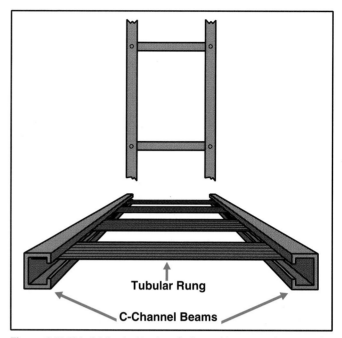

Figure 2.50 This folding ladder has C-channel beams and rectangular tubular rungs.

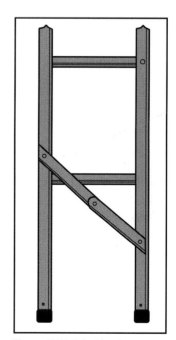

Figure 2.51 A locking brace on a metal folding ladder.

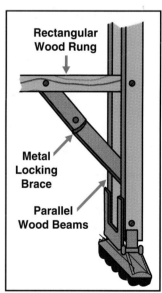

Figure 2.52 Each beam of the wood folding ladder is actually two parallel wood beams separated only by the width of the rectangular-shaped wood rungs.

noted in the following sections, is the same as that for single, roof, extension, and pole ladders. Four variations of design are found; all provide an A-frame in combination with some other type ladder.

COMBINATION SINGLE/A-FRAME LADDER

The combination single/A-frame ladder consists of two equal length ladder sections joined together at the ends of the beams by steel hinges.

The hinges lock open at the 180-degree position to form a single ladder. When used as an A-frame, two manually operated metal braces keep the two halves at the proper angle and prevent them from spreading apart when a load is applied (Figure 2.53).

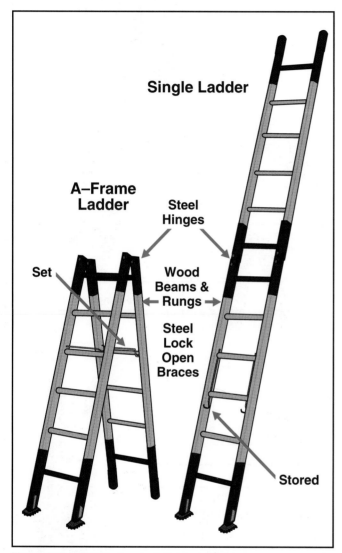

Figure 2.53 A combination single/A-frame ladder.

TELESCOPING BEAMS: COMBINATION SINGLE/A-FRAME LADDER

This ladder is constructed of metal only. There are two equal-length ladder sections joined together at the ends of the beams by steel hinges. The hinges have two locking positions. One locking position is at the 180-degree position; this position is used to form a single ladder. The second locking position, which is used instead of braces between sections, maintains the proper distance between the two halves and prevents them from spreading apart when used in the A-frame configuration.

The lower half of each section has U-channel beams that curve outward toward the butt. This feature makes the ladder wider at the bottom than in the middle or top portions. Square rungs are attached to the top of the beam. This design allows the upper half of each section to slide into the U-channel of the lower half. The beams of the upper half of the ladder are extruded rectangular channels. The rungs are hollow rectangular units. Latching devices are attached to each side of the upper rung on the lower half of the ladder. These latching devices insert into the hollow end of the rungs on the upper half of the ladder. This procedure secures the ladder at the desired height (Figure 2.54).

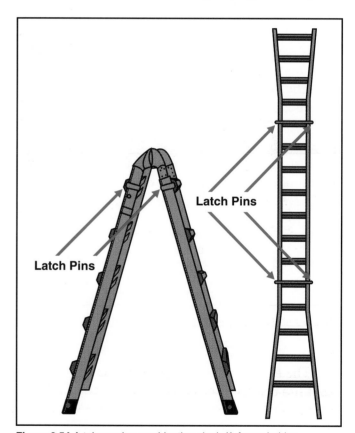

Figure 2.54 A telescoping combination single/A-frame ladder.

COMBINATION EXTENSION/A-FRAME LADDER

This ladder has no halyard and is basically a short extension ladder that either spreads apart or is spread apart to form an A-frame. Some means of joining the two sections together at the top of each section is provided so that an A-frame can be made. One type has a slot in the beam at the tip end of the bed section and a rod that protrudes from the side of the beam at the tip end of the fly section. The rod fits into the slot to make the A-frame (Figure 2.55).

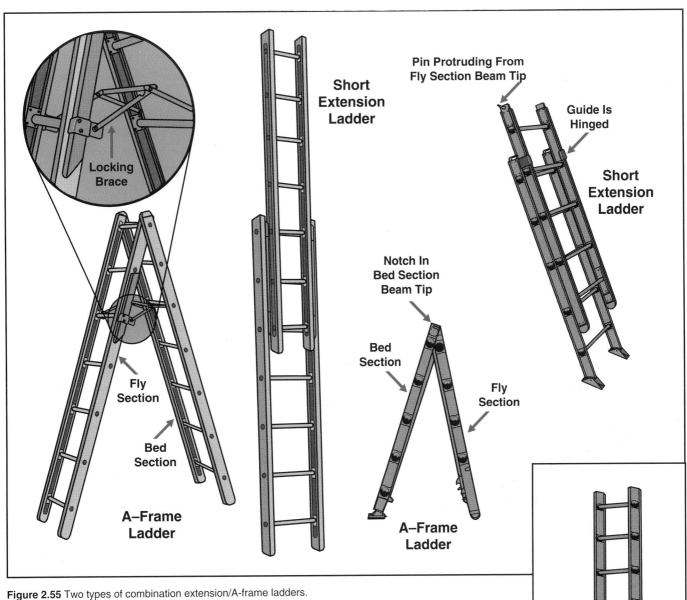

Figure 2.55 Two types of combination extension/A-frame ladders.

EXTENDING A-FRAME LADDER

The extending A-frame ladder has a fly section that fits between the two A-frame sections. A steel hinge holds the two sections of the A-frame together, and it is locked when the sections are spread open to hold them in the proper position. The hinge is made in such a way that the two A-frame sections are held far enough apart at the top that the fly section can be extended between them. There is no halyard. A modified pawl assembly holds the fly at the desired height (Figure 2.56).

Pompier Ladders

Early pompier ladders were manufactured of wood; however, most of those made today are of metal construction. NFPA 1931 restricts the length of new pompier ladders to 16 feet (5 m); however, older models may be found up to 20 feet (6 m) in length.

Metal pompier ladders have a single beam made of aluminum alloy. The beam is drilled so that the aluminum alloy rung passes through it. The rung is attached to the beam on

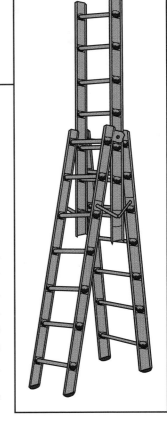

Figure 2.56 An extending A-frame ladder.

both sides by an L-shaped bracket that is riveted in place. The minimum overall width of rungs is 12 inches (305 mm).

A minimum of two metal standoff brackets are attached to the beam (one near the top and one near the bottom) on the side that will be against the building. Longer ladders may have additional standoff brackets in the middle. The standoff brackets set the ladder away from the building a minimum distance of 7 inches (180 mm) so that the climber can get both a toehold and a handgrip. A serrated gooseneck-shaped hook made of hardened steel is attached to the top of the beam. Its purpose is to hook over a windowsill or similar surface to keep the ladder from slipping (Figure 2.57).

Wood pompier ladders have a single solid-wood beam, usually of hickory. Wood rungs are supported and attached by steel L-brackets. The standoff brackets and gooseneck are much the same as those on metal ladders.

HARDWARE AND ACCESSORIES

Obviously, there is more to a ladder than beams and rungs. There are many other parts of the ladder that make it usable. In the following sections, we will look at all the different types of hardware and accessories that are used on various types of ladders. Hardware and accessories have to be designed so that they are at least as strong as the rest of the ladder. It is also important that these pieces be resistant to corrosion.

Pawls (Dogs)

Pawls, sometimes referred to as dogs, are used on fly sections of extension ladders to hold the fly sections at the desired height. There are two basic types of pawls: enclosed automatic latching and manual latching.

ENCLOSED AUTOMATIC LATCHING PAWLS

Enclosed automatic pawls consist of a metal housing and assembly that is bolted to the inside of the fly section beam, either at the bottom rung or at the next to the bottom rung. One pawl is mounted on each beam of the fly section(s). The pawls also support the rung located at this point (Figure 2.58).

There are three main parts to this pawl: the hook, the finger, and the torsion spring. The locked position is the normal position during climbing and nesting (Figure 2.59). When the fly is raised slightly, the lower section rung makes contact with the projecting (tapered) end of the finger, depressing it

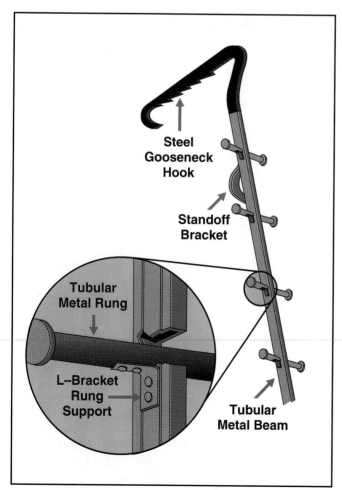

Steel Gooseneck Hook

Standoff Bracket

Tubular Metal Rung

L–Bracket Rung Support

Tubular Metal Beam

Figure 2.57 Metal pompier ladder construction.

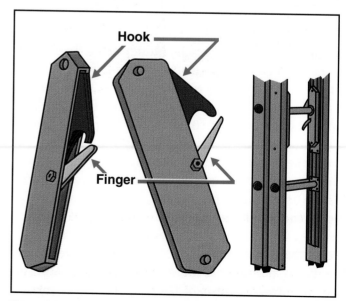

Hook

Finger

Figure 2.58 Enclosed automatic locking pawls.

into the assembly housing (Figure 2.60). When the other (rounded) end of the finger is rotated by this movement, it depresses the hook part way, putting the spring into tension. As soon as the finger clears the rung, the spring pushes the hook out again; this movement rotates the rounded end of the finger and brings the tapered end of the finger out again.

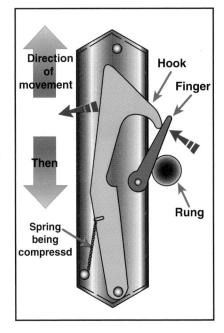

Figure 2.61 When the fly is lowered, it must first be raised a short distance to allow the finger to get above the rung. Then, as it is lowered, the finger begins to depress the hook so that it will clear the rung.

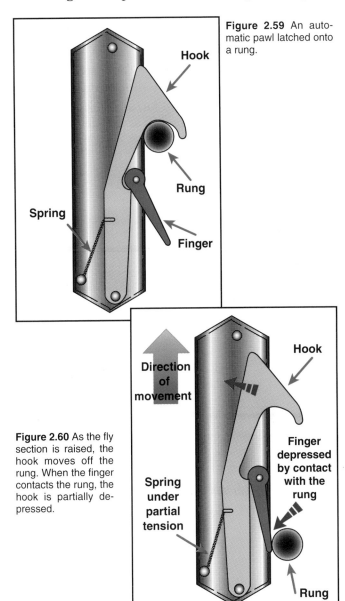

Figure 2.59 An automatic pawl latched onto a rung.

Figure 2.60 As the fly section is raised, the hook moves off the rung. When the finger contacts the rung, the hook is partially depressed.

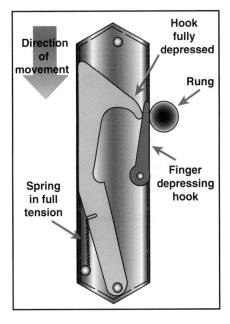

Figure 2.62 As the fly continues downward, the finger depresses the hook into the pawl housing so that it will clear the rung.

When the slanted portion of the top of the hook contacts the bottom of the next higher rung, it again depresses the assembly into the housing. The hook slides by the rung, then immediately springs out; it can be latched at this point. To lower a fly section, the fly is extended until the hooks and the fingers clear the rung (Figure 2.61). The fly is then lowered until the fingers depress the hooks (Figure 2.62).

MANUAL LATCHING PAWLS

The manual latching pawl uses two A-shaped steel pawls attached near each end of a steel rod that runs between the beams approximately 4 inches (100 mm) above the bottom rung of the fly section (Figure 2.63). This rod is mounted so that it will rotate. Rotation of the rod moves the pawls in and out from between the rungs of the bed section. A continuous halyard (described below) is used with manual latching pawls. Both ends of the halyard are snapped to an inverted, L-shaped halyard anchor that is attached to the center of the steel rod. The end of the halyard that comes down

Figure 2.63 A manual latching pawl. *Courtesy of ALACO Ladder Company.*

from the pulley is snapped to the short leg of the inverted L part. The end that comes up from the butt of the ladder is snapped to the long leg of the inverted L part (Figure 2.64).

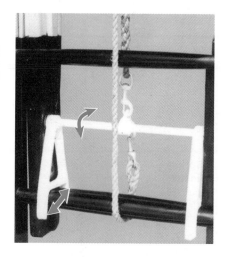

Figure 2.64 A manual pawl in the latched position. *Courtesy of ALACO Ladder Company.*

When the halyard is operated in the normal manner to raise the fly, pressure is applied to the short leg of the L part of the halyard anchor. As soon as the pawls clear the bed section rung, the rod rotates clockwise, and the pawls are pulled away from the bed section. This action keeps them clear of bed section rungs as the fly travels upward (Figure 2.65).

To latch or set the pawls, the fly is raised until the pawls are slightly above the bed section rung upon which they are to rest. Tension is maintained on the halyard while one hand is used to pull upward on the slack or lower part of the halyard rope (the part that runs down around the bottom rung of the bed section and back up to the long leg

of the L part of the halyard anchor). This action causes the rod to rotate counterclockwise and the pawls are pulled back up between the rungs of the bed section. The fly section is then eased downward until the pawls engage the bed section rung (Figures 2.66 a and b).

Halyard/Halyard Anchor/Pulley

The halyard rope is used to extend the fly sections of an extension ladder. NFPA 1931 specifies that the halyard rope be a minimum of ⅜-inch (10 mm) in diameter with minimum breaking strength of 825 pounds (374 kg) and be of sufficient

Figure 2.65 Pulling downward on the halyard (1), lifts the fly (2), and allows the pawl to swing back (3). *Courtesy of ALACO Ladder Company.*

Figure 2.66a When the desired height is attained, tension is maintained on the halyard (1), while the other part of the halyard (2) is pulled with the other hand. *Courtesy of ALACO Ladder Company.*

Figure 2.66b The pawl swings outward above the rung upon which it will rest. Tension is maintained on the halyard (3) while the first part (4) is slacked off to allow the fly to lower and the pawl to come to rest on the rung. *Courtesy of ALACO Ladder Company.*

length for the purpose intended. The halyard rope is threaded through a pulley attached to the top rung of the bed section. One end is attached to the bottom rung of the fly section. The other end is either free, in which case it is known as a *free end halyard,* or it runs down the ladder, under the bottom rung, and back up to the bottom rung of the fly section. This type is called a *continuous halyard.*

Three- and four-section extension ladders have a second halyard, usually a cable, that threads through a pulley attached to the top rung of the intermediate fly section(s). One end is attached to the bottom rung of the top fly section. The other end is attached to the second rung from the tip of the bed section (Figure 2.67). In this case, the second halyard is most commonly a wire cable. The cable must be ³⁄₁₆-inch (5 mm) in diameter to meet NFPA 1931 specifications. Where cable is used, a means for adjusting the length of cable has to be provided. No splices are allowed.

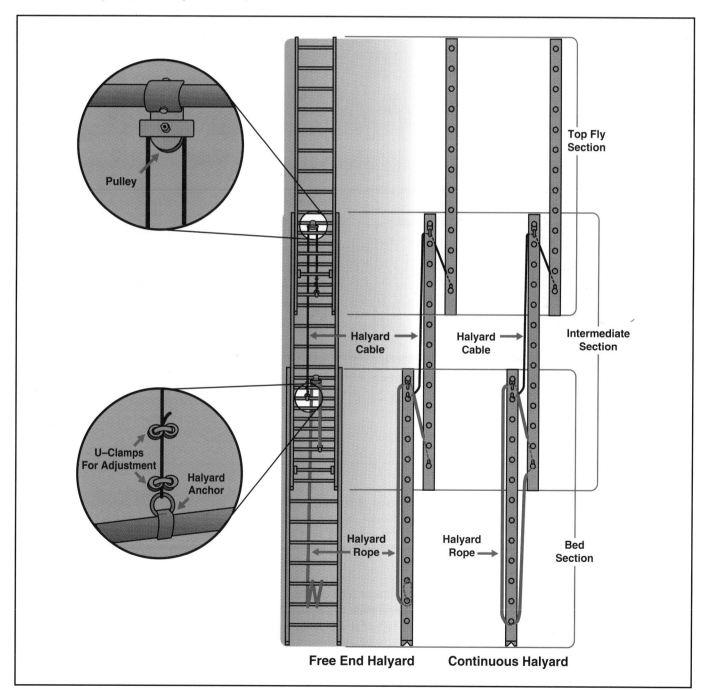

Figure 2.67 A typical halyard arrangement for a three-section extension ladder.

Halyard anchors are used only on ladders that contain a free end halyard. The anchors are attached to the bottom rung of each fly section. These anchors consist of a metal band that goes around the rung and a small metal ring, to which the rope or cable is attached (Figure 2.68). The pulleys are attached in a similar manner to the halyard anchors. However, the pulleys are attached to the top rung of the bed section and each intermediate fly section (Figure 2.69).

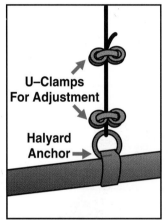

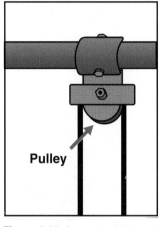

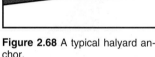

Figure 2.68 A typical halyard anchor.

Figure 2.69 A standard halyard pulley arrangement.

Stops

All extension ladders have stops to keep fly sections from extending off the top end of the ladder. One type consists of both an L-shaped piece of metal attached to the outside of the fly section beam near its bottom and a similar piece of metal attached to the top of the bed section beam near its top (Figure 2.70). As the fly extends upward, the two pieces engage and prevent further extension of

the fly(s). Another type uses blocks set in the guide tracks. A third type uses attachments on an upper rung that stop the pawls from extending beyond that point (Figure 2.71).

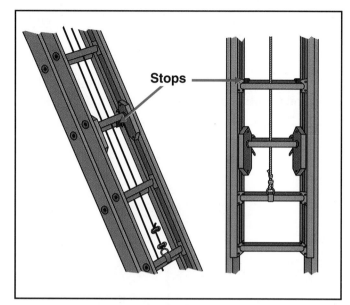

Figure 2.71 Other types of ladders use a ring attached to each end of a rung.

Butt Spurs

Butt spurs are intended to prevent the ladder butt from slipping. They are most effective on soft surfaces. As with other features of ladder construction, there are a number of different designs used for the butt spurs. The type of butt spur used on any particular ladder depends on the design of the ladder. There are different butt spur arrangements for wood, metal, and fiberglass ladders (Figures 2.72 a through c). All butt spurs are constructed of metal, usually steel.

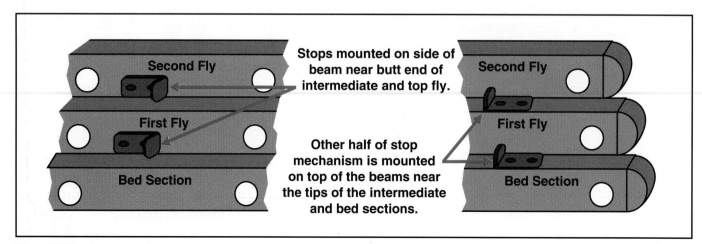

Figure 2.70 Common types of stops used for extension and pole ladders.

Figure 2.72a Butt spur on a wood ladder.

Butt Spurs

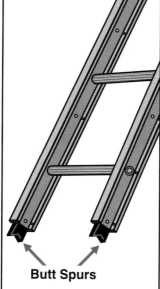

Butt Spurs

Figure 2.72b Butt spur on a solid beam metal ladder.

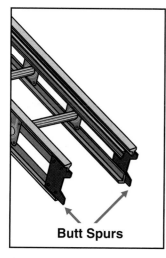

Butt Spurs

Figure 2.72c Butt spur on a truss beam metal ladder.

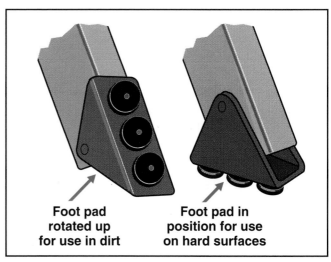

Foot pad rotated up for use in dirt   **Foot pad in position for use on hard surfaces**

Figure 2.73 Foot pads are useful on hard, flat surfaces.

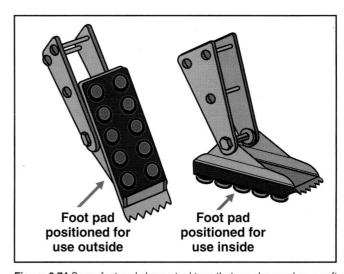

Foot pad positioned for use outside   **Foot pad positioned for use inside**

Figure 2.74 Some foot pads have steel toes that may be used on a soft surface.

Figure 2.75 Steel tie rods are used to help hold wood ladders together.

Foot Pads

Foot pads are also intended to prevent the ladder butt from slipping. Foot pads are most effective on hard surfaces. Foot pads are made of metal with rubber or rubberlike treads. They are either bolted or riveted to the butt of the beam, usually on shorter length ladders, particularly on folding ladders (Figure 2.73). Some are manufactured with a combination rubber tread and steel toe for use either inside or outside a building (Figure 2.74).

Tie Rods

Wood ladders have steel tie rods to help hold them together. Tie rods are usually installed just below every fourth rung. The head and nut are flat, and the outside of the beam is countersunk so that they do not protrude (Figure 2.75). The nut is notched so that a two-pronged tool can be used to tighten it.

Toe Rod

A *toe rod* is a steel rod that is installed just above the butt spurs. It provides a place for the firefighter to place the toe of the boot during heeling operations (Figure 2.76). The toe rod has an appearance similar to that of a rung, but it is not intended as a step.

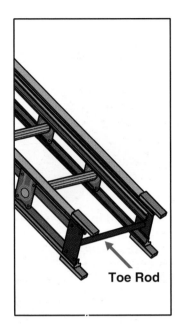

Figure 2.76 A toe rod installed at the butt of a ladder.

Toe Rod

Figure 2.78 Staypole spurs help keep the pole in place once it is positioned.

Staypoles, Staypole Spurs, and Toggles

Staypoles are used both to raise and support ground ladders that are 40 feet (12 m) or taller. The staypoles are most typically constructed of the same material as the beams of the ladder to which they are attached (wood, metal, or fiberglass). NFPA 1931 requires staypoles to be permanently attached to the beams of the ladders. However, models built before this attachment was required by the standard may be equipped with detachable staypoles. Regardless of which type the ladder contains, the staypoles are all attached to the side of the beam near the top of the bed ladder. The attaching device is called a *toggle* (Figure 2.77). Metal spurs are provided on the ends of the staypoles to prevent slippage when the ladder is in place and the poles have been set (Figure 2.78).

Figure 2.77 A typical staypole toggle.

Mud Guard

A mud guard is an accessory available for metal truss ladders. It is a metal plate attached between the beams at the butt, which significantly increases the surface area resting on the ground. Its purpose is to prevent the ladder butt from sinking into soft ground (Figure 2.79).

Figure 2.79 A mud guard installed across the butt of a ladder.

Protection Plates

Protection plates are strips of metal attached to ladders at chafing points, such as the tip, or at areas in contact with the apparatus mounting brackets (Figure 2.80). They are intended to reduce any damage that might occur either when loading or unloading the ladders or from vibrations due to normal road travel.

Levelers

Fire department ground ladders (with the exception of roof and pompier ladders) are designed to be used on level surfaces. Loads and stresses are thereby distributed evenly between the two beams.

When ground ladders have to be placed on uneven terrain, steps must be taken to either shim or otherwise support the beam that is not touching the supporting surface. A device known as a leveler has been developed to overcome this problem. Levelers are used where uneven terrain is regularly encountered.

There are several types of levelers available. One type is permanently attached to the butt. This type automatically adjusts to terrain contours (Figure 2.81). A second type of leveler is carried on the apparatus and is attached as necessary (Figure 2.82). It is manually adjusted by the firefighters who place the ladder.

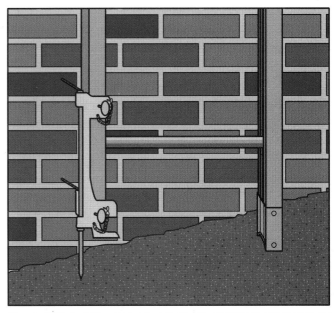

Figure 2.82 Portable, manually adjusted leveler that is attached to the ladder beam as needed.

Roof Ladder Hooks

Hooks for roof ladders are constructed of solid steel. The end of the hook intended to engage the roof is tapered to a point to reduce slippage. The base of the hook is square where it passes through the top of the spring housing. The top opening in the spring housing is also square. This design prevents the hook from rotating unless it is depressed and allows it to lock into two positions, one with the hook nested between the rungs and the other with the hook parallel to the beam (90 degrees from the rung). A coil spring inside the housing holds the hook in the desired position (Figure 2.83). To change the position, the hook is manually depressed until the square part clears the hole in the housing; then it is rotated.

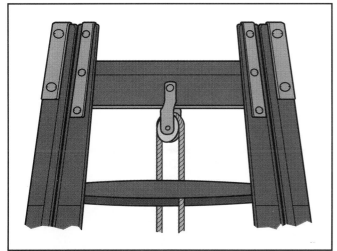

Figure 2.80 Protection plates are used on the tip ends of the beams to prevent scuff damage.

Figure 2.81 Permanently attached automatic adjusting leveler.

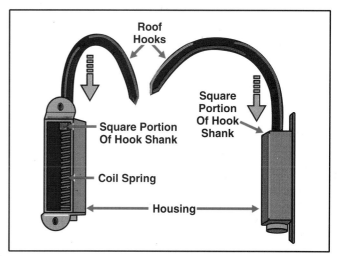

Figure 2.83 A common roof ladder hook assembly.

Labels

There are a number of labels found on fire service ground ladders. With revisions to the NFPA 1931 standard over the years, labeling requirements for ladders have evolved. The labels listed below are those required by the 1994 edition of the NFPA 1931. Bear in mind that older ladders may not have all of these labels. In most cases, should the fire department wish to, newer labels may be acquired from the manufacturer and placed on older ladders. Labels that are commonly found are as follows:

- Manufacturer's Identification — This label gives the manufacturer's name and address and identifies the series or model number of the ladder (Figure 2.84).

- Serial Number — NFPA 1931 requires that fire department ground ladders bear a unique individual identification number. It can be embossed, stenciled, branded, or stamped on the ladder, or it may be on a metal plate attached to the ladder (Figure 2.85).

- Certification Label — This label attests that the ladder has been manufactured in accordance with NFPA 1931 and U.S. Occupational Safety and Health Administration (OSHA) fire ladder requirements (Figure 2.86).

- Heat Sensor Label — These are labels that are affixed to the inside of each beam of each ladder section. One label is attached immediately below the second rung from the tip of each section and another immediately below the center rung of that section. The sensor should turn a different color at 300°F (149°C), plus or minus 5 percent. The color change indicates that the ladder has been exposed to a sufficient degree of heat that it should be tested before further use (Figures 2.87 a and b).

- Electrical Hazard Warning Label — This label is intended to remind firefighters of the dangers of coming in contact with electrical wires. The labels vary slightly depending on whether they are placed on a wood, fiberglass, or metal ladder. The primary difference is that metal ladders must carry the additional warning that the ladder will conduct electricity (Figures 2.88 a and b). These labels must be placed on the outside of both beams, between 4½ and 6 feet (1.4 m and 1.8 m) from the butt of the ladder

- Ladder Positioning Label — This label is intended to remind the firefighter of the proper placement of the ladder. Included are directions for placing the ladder at a proper climbing angle and to which direction the fly sections of the ladder should be facing (Figure 2.89). These labels must be placed on the outside of both beams, between 4 ½ and 6 feet (1.4 m and 1.8 m) from the butt of the ladder.

Any labels that are not clearly legible should be replaced as soon as possible.

Figure 2.84 The ladder manufacturer will place its label on each ladder.

Figure 2.85 The serial number should be plainly visible somewhere on the ladder beam.

Figure 2.86 The certification label assures the user that the ladder meets NFPA requirements.

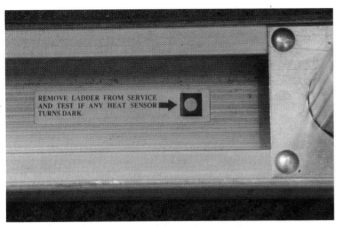

Figure 2.87a A heat sensor label that includes directions on how to read it.

Figure 2.87b Some heat sensor labels do not contain directions.

Figure 2.88a An electrical hazard label for metal ladders.

Figure 2.88b An electrical hazard label for wood ladders.

CAUTION

Set Up Ladder Properly
To Reduce Slip And
Overload Hazards.
Follow These Instructions.

75°
(approx.)

① Place Toes Against Bottom Of Ladder Siderails.

② Stand Erect.

③ Extend Arms Straight Out.

④ Palms Of Hands Should Touch Top Of Rung At Shoulder Level.

OUT

Figure 2.89 This label shows in which direction the fly should be facing, and helps firefighters assure that the ladder is at a proper climbing angle.

INSPECTION AND MAINTENANCE OF GROUND LADDERS

Like any equipment used in the fire service, ground ladders require periodic and thorough inspection, regular cleaning, and lubrication to ensure that they are safe and 100 percent operational. This degree of reliability does not occur by accident. A fire department must follow a systematic program to accomplish this. NFPA 1932, *Standard on Use, Maintenance, and Service Testing of Fire Department Ground Ladders* mandates a specific program for fire departments to check their ground ladders.

Inspection

NFPA 1932 requires ladders to be inspected after each use and on a monthly basis. Some of the things that should be checked on all types of ladders include the following:

- The heat sensor label on metal and fiberglass ladders. Check to see if there is a color change indicating heat exposure. Ladders without a heat sensor label may also show signs of heat exposure such as bubbled or blackened varnish on wood ladders, discoloration of fiberglass ladders, heavy soot deposits or bubbled paint on tips of any kind of ladder. (CAUTION: Metal ladders which have been exposed to heat shall be placed out of service until tested.)

- Rungs for snugness and tightness (Figure 2.90).

- Bolts and rivets for tightness (bolts on wood ladders should not be so tight that they crush the wood).

- Welds for any cracks or apparent defects.

- Beams and rungs for cracks, splintering, breaks, gouges, checks, wavy conditions, or deformation.

In addition to these general things, there some other items that need to be checked, depending on the specific type of ladder being inspected. The following sections highlight some of these items.

WOOD LADDERS/WOOD COMPONENTS OF COMPOSITE LADDERS

The following things must be examined on wood ladders or composite ladders with wood components:

- Look for areas where the varnish finish has been chafed or scraped off.

- Check for darkening of the varnish (indicating exposure to heat).

- Check for dark streaks in the wood (indicating deterioration of the wood). (CAUTION: Any indication of deterioration of the wood is cause for the ladder to be removed from service until it can be service tested.)

ROOF LADDERS

Make sure that the roof hook assemblies operate with relative ease (Figure 2.91). In addition, the assembly should not show signs of rust, the hooks should not be deformed, and parts should be firmly attached with no sign of looseness. (NOTE: Serious

Figure 2.90 Check the rungs to make sure they are tight.

Figure 2.91 The roof hooks should fully open with ease.

problems found should result in removal of the ladder from service pending service testing.)

EXTENSION AND POLE LADDERS

The following must be checked on extension and pole ladders:

- Make sure the pawl assemblies work properly. The hook and finger should move in and out freely.

- Look for fraying or kinking of the halyard (Figure 2.92). If this condition is found, the halyard should be replaced.

Figure 2.92 Run the halyard through your fingers to check for frays or cuts.

- Check the snugness of the halyard cable when the ladder is in the bedded position. This check ensures proper synchronization of the upper sections during operation.

- Make sure the pulleys turn freely.

- Check the condition of the ladder guides and for free movement of the fly sections.

- Check for free operation of the staypole toggles and check their condition. Detachable staypoles are provided with a

latching mechanism at the toggle. This mechanism should be checked to be sure that it is latching properly.

If any of the conditions described in the preceding items are found, the ladder should be removed from service until it can be repaired and tested. Ladders that cannot be safely repaired will have to be destroyed or scrapped for parts.

> ### WARNING
> **Failure to remove a defective ladder from service can result in a dramatic ladder failure that injures or kills firefighters.**

Maintenance

Before discussing ladder maintenance, it is important to understand the difference between maintenance and repair. The word *maintenance,* as used here, means keeping ladders in a state of usefulness or readiness. *Repair* means to either restore or replace that which has become inoperable. All firefighters should be capable of performing routine maintenance functions on ground ladders. Any ladders in need of repair will require the service of a trained ladder repair technician.

NFPA 1932 lists the following general maintenance items that apply to all types of ground ladders:

- Keep ground ladders free of moisture.

- Do not store or rest ladders in a position where they are subjected to exhaust or engine heat (Figure 2.93).

Figure 2.93 Never lean the ladder near the apparatus exhaust pipe.

- Do not store ladders in an area where they are exposed to the elements.

- Ladders should not be painted, except for the top and bottom 12 inches (300 mm) of the beams for purposes of identification or visibility.

WOOD LADDERS/WOOD COMPONENTS OF COMPOSITE LADDERS

There are some maintenance requirements that are specific to wood ladders only:

- Store wood ladders away from steam pipes, away from radiators, and out of direct sunlight or areas where the humidity is artificially reduced. Continued exposure to either a heating source or direct sunlight may cause the wood to dry out and lose its strength.

- Wood ladders that come from the factory with three coats of phenolic-tung oil varnish with ultraviolet (UV) ray absorbers should be spot refinished as needed, such as when scratches, chips, gouges, or scorching are observed. Use any phenolic-tung oil product with UV absorbers, also known as marine spar varnish with sunblockers, to accomplish this task.

- Approximately every three years, when the finish either loses its glossy appearance or becomes dull, apply one coat of phenolic-tung oil varnish with UV ray absorbers to the entire ladder.

The procedure for spot refinishing or repairing small areas of damage to the varnish coating are as follows:

Step 1: Clean damaged area(s) by sanding to remove all loose or damaged wood. Finish sand with fine paper, feathering into the surrounding undamaged finish (Figure 2.94).

Step 2: Apply one coat of phenolic-tung oil varnish with UV absorbers; one part thinner or turpentine to three parts varnish. This is a primer or sealer coat.

Step 3: Before the sealer coat is fully dry, apply a second coat as it comes from the can (Figure 2.95). Before that coat is fully dry, apply a final coat from the can. No sanding between coats is necessary.

Figure 2.94 Sand areas of the ladder that will receive lacquer.

Figure 2.95 Lacquer may be applied with a rag or brush.

METAL AND FIBERGLASS LADDERS

Metal ladders simply require an occasional application of a good automotive paste wax to preserve the surface finish. Fiberglass ladders may require an application of urethane varnish to restore the surface if it has been adversely affected by the sun's ultraviolet rays. Ultraviolet rays may cause the resins in the fiberglass to deteriorate. The following procedure may be used to renew the surface coating:

Step 1: Clean the surface thoroughly with paint thinner, removing all loose material. Let dry. Sand lightly if necessary.

Step 2: Apply a coat of urethane varnish. Let dry thoroughly. Sand lightly.

Step 3: Add additional coats of varnish, as needed, to result in a smooth, glossy surface.

ROOF LADDERS

The major portions of roof ladders are maintained the same as any other ladder. The only portions that require special attention are the hooks. When roof ladder hooks are rusted in or around the spring assembly housing, they should be disassembled, cleaned, and lubricated.

EXTENSION AND POLE LADDERS

The following maintenance items are specific to extension and pole ladders:

- Pawl assemblies should be kept clean and lubricated in accordance with manufacturer's instructions.

- Pawl torsion springs should be replaced every five years or sooner if pawl operation appears weak. When reinstalling pawl assemblies, use caution to prevent overtightening of pawl assembly fasteners. Overtightening will cause binding of pawl assembly parts.

- Ladder slide areas (including guides) should be kept lubricated in accordance with manufacturer's instructions.

- When replacing halyards on a three- or four-section extension ladder, it may be wise to sketch the cable arrangement before removing the old halyard. The sketch can then be used to ensure that the replacement is threaded properly.

- In order for halyard pulleys with ball bearing centers to operate smoothly, a small amount of lubricant should be applied periodically.

Cleaning Ladders

Regular and proper cleaning of ladders is more than a matter of appearance. Unremoved dirt or debris may collect and harden to the point where ladder sections are no longer operable. Therefore, it is recommended that ladders be cleaned after every use.

A brush and running water are the most effective tools for cleaning ladders (Figure 2.96). Tar, oil, or greasy residues should be removed with safety solvents. After the ladder is rinsed, or anytime a ladder is wet, it should be wiped dry. During each cleaning period, firefighters should look for defects. Any defects should be handled through local fire department procedures.

Figure 2.96 The ladder should be cleaned of fire scene and road grime.

Repairing Ladders

During the inspection of ladders, the inspector should mark all defects with chalk or some other suitable marker (Figure 2.97). Legible marks help keep defects from being overlooked when repairs are made. All ladder repairs must be made in accordance with manufacturer's recommendations. These ladders must be tested according to NFPA 1932 prior to being placed back into service.

Minor repairs to wood ladders can usually be made in fire department shops. Special care must be observed when repairing wood ladders. Small splinters can be repaired by cutting the wood across the grain with a sharp knife at the large end or base of the splinter. The splinter can then be removed and the part sanded smooth. When tightening wood ladder beam bolts and tie rods, a special spanner should be used, and the nut should only be tightened snugly to prevent crushing the wood cells.

Figure 2.98a Some departments choose to paint the ends of their ladders.

Figure 2.98b Painting the racking point on a ladder speeds the loading process.

Figure 2.97 Defects should be marked on a ladder and then the ladder taken out of service. If no other markers are available, fire line tape may be used.

The finish of the wood is very important in maintaining the maximum useful life of the ladder. As a wood finish decomposes or oxidizes, the moisture, resins, and oils in the wood escape and leave the wood cells exposed to air and to the elements. Varnish may be depended upon to preserve the wood, because it seals in the natural oils and resins and keeps moisture out. Varnish also prevents dry rot and fungus growth. When the varnish finish becomes worn or scratched, it should be repaired without delay. Paint is not recommended on fire department ladders except to identify the ladder ends, balance point, racking points, or length (Figures 2.98 a and b). Paint used as an outer cover for a wood ladder makes it practically impossible to detect fungus growth, dry rot, or cracks during inspection.

The preparation of the wood to be refinished and the refinishing procedure are of prime importance. The department should check with the manufacturer of the ladder, and comply with the manufacturer's suggested practices. First, remove as much of the old finish as possible by scraping with a scraper tool. (NOTE: Do not scrape against the grain of the wood.) Sand all surfaces with medium sandpaper or garnet paper; follow with fine sandpaper. An appropriate liquid remover may be used when conditions warrant. To preserve the wood, apply two coats of sealer. Allow sufficient drying time between coats. After these applications, apply two coats of varnish, allowing sufficient drying time between coats. All repairs should be recorded on the correct ladder report form.

TESTING GROUND LADDERS

If one reviews all of the possible events that may require use of fire service ground ladders, a conclusion can easily be made that fire service ladders are subject to abuse and overloading. There are two primary types of testing that ground ladders must endure: design verification testing and service testing. The remainder of this chapter is dedicated to the testing of ground ladders.

Design Verification Testing

NFPA 1931 requires ladder manufacturers to conduct design verification tests as a part of the initial evaluation of a specific product design. Tests must again be conducted when there is a change in the design, method of manufacturing, or material. Note that each and every ladder manufactured is NOT tested in this manner, only each new model is so tested. The list of design verification tests is extensive. Ladders that are used for design verification testing are destroyed after testing is completed.

Fire service personnel need not worry about the specifics of design verification testing. Fire service personnel are never involved in this testing. Fire service personnel only need to be aware that the ladders they purchase have met this requirement.

Service Testing

No manufacturer or fire department official can guarantee that a ladder will not fail during use, but the chances can be minimized by testing them in accordance with NFPA 1932. This standard recommends that only the tests specified be conducted — either by the fire department or by an approved testing organization. Caution must be used when performing service tests on ground ladders to prevent damage to the ladder or injury to personnel.

WARNING
Personnel need to be aware that there is a possibility of sudden failure of the ground ladder undergoing service testing. All safety precautions possible should be taken to avoid injury.

Service tests of ground ladders may require the purchase of measuring instruments. Testing should be performed only by personnel trained in service test procedures and in the operation of the service test equipment.

TESTING FREQUENCY

NFPA 1932 contains the following requirements pertaining to frequency of service testing:

- At least annually
- Anytime a ladder is suspected of being unsafe
- After the ladder has been subjected to overloading
- After the ladder has been subjected to impact loading or unusual conditions of use
- After heat exposure (NOTE: Metal ground ladders being tested because of exposure to heat may be subjected to either the Strength Service Test or the Hardness Service Test; both are not required. However, if the Hardness Service Test is used and the ladder fails, a Strength Service Test shall be conducted.)
- After any deficiencies have been repaired, unless the only repair was replacing the halyard

Any signs of failure during service testing is sufficient cause for the ground ladder to be removed from service. It should either be repaired and retested or be destroyed. The following sections highlight the service tests that should be performed on ground ladders.

STRENGTH TESTS FOR ALL LADDERS, EXCEPT POMPIER AND FOLDING LADDERS

The following tests are designed to ensure that the ladder has adequate strength to be used safely.

Horizontal Bending Test

All ladders should be subjected to the horizontal bending test. The following procedure may be used:

Step 1:   Rest the ladder on 1-inch (25 mm) cylindrical supports that are placed 6 inches (150 mm) from each end (Figure 2.99). These supports must be high enough off the ground to allow for normal ladder deflection (sagging). If the ladder is an extension or combination ladder, it is fully extended.

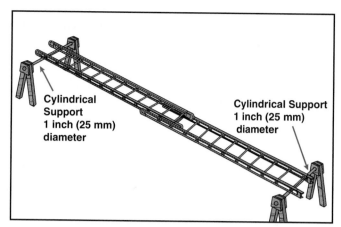

Figure 2.99 Support the ladder on 1-inch (25 mm) rods.

Step 2: Locate the test load area, which is a space 32 inches (800 mm) long over the actual center of the ladder. Place over the test load area some type of board or a plate on which the free weights can be set (Figure 2.100). Remember to include the weight of the board or plate as part of the overall test weight.

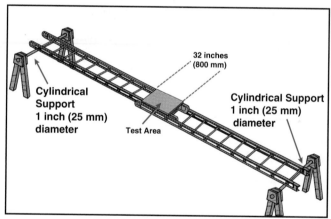

Figure 2.100 The test area is in the center of the ladder.

Step 3: Add the preload test weights slowly until 350 pounds (159 kg) are in place (Figure 2.101). (**NOTE:** For ladders made before 1984, use 300 pounds [136 kg].) Allow the load to sit for one minute, and then remove the weights.

Step 4: Measure and record the distance from the bottom of each side rail to the ground (Figure 2.102).

Step 5: Apply the test load of 500 pounds (227 kg) to the ladder (Figure 2.103). (**NOTE:** For ladders made before 1984, use 400 pounds [181 kg].) Allow the load to sit for five minutes, and then remove the weight.

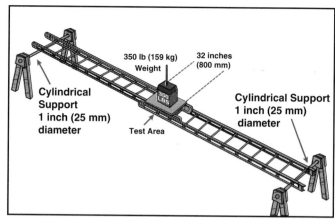

Figure 2.101 The 350 pounds (159 kg) are slowly added to the test area.

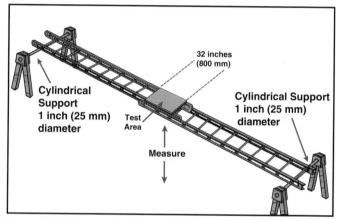

Figure 2.102 Measure the distance from the ladder to the ground.

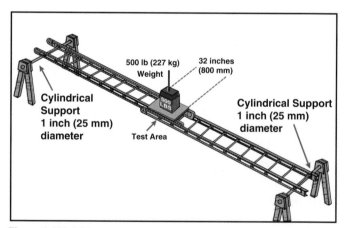

Figure 2.103 Add 500 pounds (227 kg) to the ladder.

Step 6a: *For Metal and Fiberglass Ladders* — Five minutes after the load has been removed, again measure the distance from the bottom of each side rail to the ground. The difference in distances should be less than those listed in Table 2.3; any ladder that does not fall within these measurements should be removed from service.

TABLE 2.3	
Horizontal Bending Test Distances	
Length of Ladder	**Difference in Measurements**
25 feet or less (7.6 m)	½ inch (13 mm)
26 feet to 34 feet (7.7 m to 10.4 m)	1 inch (25 mm)
35 feet or over (10.5 m or over)	1½ inch (38 mm)

Step 6b: *For Wood Ladders* — No measurements are taken. The ladder must simply not show signs of failure during testing.

Roof Ladder Hook Test

The following procedure should be used to ensure that the hooks on a roof ladder are structurally sound and capable of holding the weight of firefighters who work on the ladder:

Step 1: Position the ladder over the edge of a wall or platform so that it is hanging solely by the points of its hooks (Figure 2.104). The wall or platform must be high enough for the ladder to hang straight with the test weights suspended from it. A safety restraint should be used to fasten the ladder to the wall so that it will not fly away and injure people should a failure occur.

Step 2: Using webbing of appropriate size and strength, suspend the test weight of 1,000 pounds (454 kg) from the ladder for at least one minute (Figure 2.105).

Step 3: Remove the test weight and inspect the hooks for damage. The ladder passes the test if no permanent deformation is found.

Extension Ladder Hardware Test

The following test is used to ensure that ladder pawls and other components of extension ladders are strong enough to hold the weight of firefighters who will work on the ladder:

Step 1: Position the ladder against a wall so that it is at a 75 ½-degree angle. The ladder is extended a minimum of one rung's length (Figure 2.106).

Step 2: Using webbing of appropriate size and strength, suspend the test weight of

1,000 pounds (454 kg) from the ladder for at least one minute (Figure 2.107).

Step 3: Remove the test weight and inspect the ladder hardware for damage. The ladder passes the test if no permanent deformation is found.

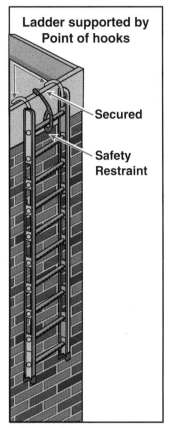

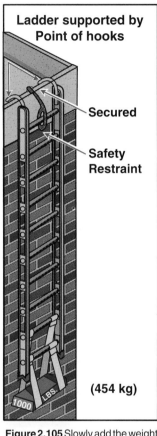

Figure 2.104 Hook the ladder over a sturdy wall.

Figure 2.105 Slowly add the weight to the ladder.

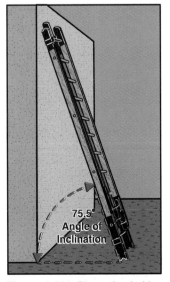

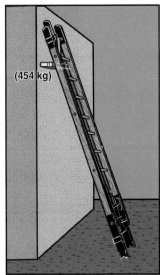

Figure 2.106 Place the ladder against the wall.

Figure 2.107 Add 1,000 pounds (454 kg) to the ladder.

STRENGTH TEST FOR POMPIER LADDERS

Use the following procedure to check the integrity of pompier ladders:

Step 1: Position the ladder over the edge of a wall or platform so that it is hanging solely by its hook. The wall or platform must be high enough for the ladder to hang straight with the test weights suspended from it.

A safety restraint should be used to fasten the ladder to the wall so that it will not fly away and injure people should a failure occur.

Step 2: Using webbing of appropriate size and strength, suspend the test weight of 1,000 pounds (454 kg) from the ladder for at least one minute (Figure 2.108).

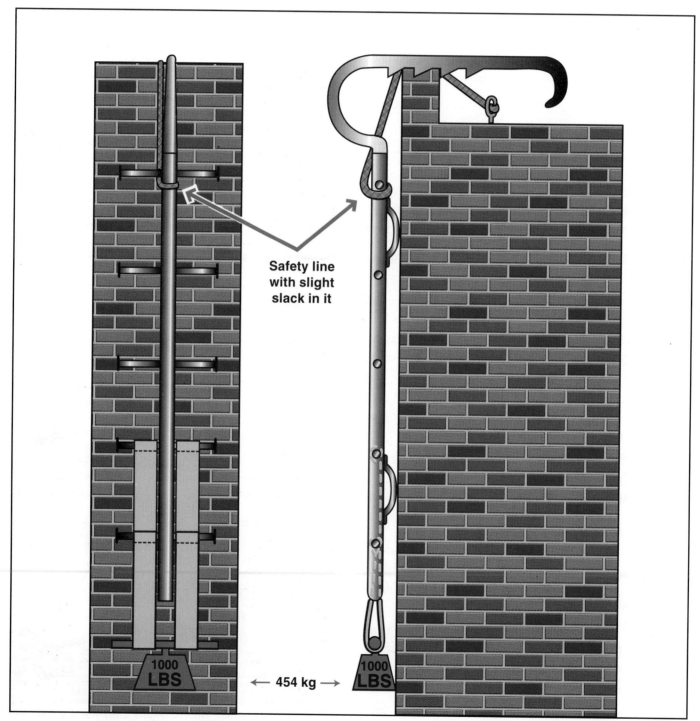

Safety line with slight slack in it

1000 LBS

← **454 kg** →

1000 LBS

Figure 2.108 The positioning and equipment needed to test a pompier ladder.

Step 3: Remove the test weight and inspect the hook for damage. The ladder passes the test if it does not ultimately fail (break).

STRENGTH TEST FOR FOLDING LADDERS

The following horizontal bending test should be used for determining the strength of folding ladders:

Step 1: Rest the ladder on 1-inch (25 mm) cylindrical supports that are placed 6 inches (150 mm) from each end (Figure 2.109). These supports must be high enough off the ground to allow for normal ladder deflection (sagging).

Step 2: Locate the test load area, which is a space 16 inches (400 mm) long over the actual center of the ladder. Place over the test area some type of board or a plate on which the free weights can be set (Figure 2.110). Remember to include the weight of the board or plate as part of the overall test weight.

NOTE: For wood folding ladders, proceed to Step 5.

Step 3: Add the preload test weights slowly until 160 pounds (73 kg) are in place (Figure

2.111). Allow the load to sit for one minute, and then remove the weights.

Step 4: Measure and record the distance from the bottom of each side rail to the ground (Figure 2.112).

Step 5: Apply the test load of 225 pounds (102 kg) to the ladder. Allow the load to sit for five minutes, and then remove the weight (Figure 2.113).

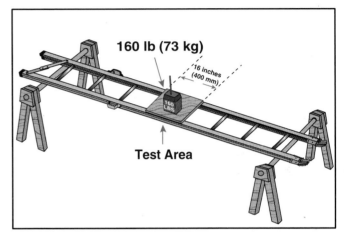

Figure 2.111 Add 160 pounds (73 kg) to the test area.

Figure 2.109 Support the folding ladder on 1-inch (25 mm) rods.

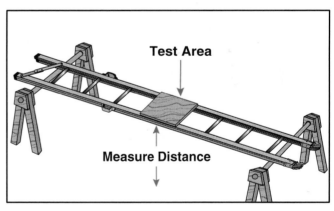

Figure 2.112 Measure the distance between the ladder and the ground.

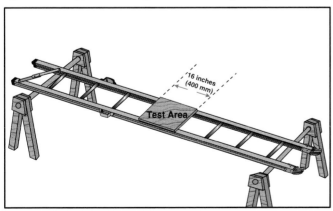

Figure 2.110 The test area is in the center of the ladder.

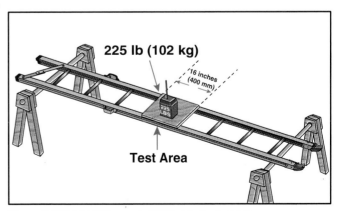

Figure 2.113 Add 225 pounds (102 kg) to the test area.

Step 6a: *For Metal and Fiberglass Ladders* — Five minutes after the load has been removed, again measure the distance from the bottom of each side rail to the ground. The difference in distances should be less than ½-inch (13 mm). Any ladder that does not fall within these measurements should be removed from service.

Step 6b: *For Wood Ladders* — No measurements are taken. The ladder must simply not show signs of failure during testing.

LIQUID PENETRANT TESTING

The liquid penetrant test is an additional nondestructive test for metal ground ladders constructed of 6061-T6 aluminum alloy. This test is usually contracted to an approved testing organization because interpretation of test results is critical and must be performed only by certified personnel.

The purpose of liquid penetrant testing is to detect otherwise invisible cracks in the metal. It is particularly valuable to check welds. For an explanation of this test, see the appendix to NFPA 1932.

RECORDS

It is important that records be kept on strength service tests, repairs, and retesting. These records will be valuable for the following reasons:

- They provide a means of showing that testing was performed.
- They provide the information necessary (dates) to assure that testing is done annually.
- They identify the service status (in or out of service).
- They identify the ladder location.
- They identify ladders that have been subjected to abuse and show that they were tested after the incident.
- The information is available for use in legal actions resulting from ladder failures.
- They are available for use in ISO inspections.
- They provide information for evaluation of one model or brand as compared with another model or brand when drawing specifications for purchasing or when evaluating purchasing bids.

Appendix A provides a sample testing and repair record sheet for fire department ground ladders.

Chapter 2 Review

Directions

The following activities are designed to help you comprehend and apply the information in Chapter 2 of **Fire Service Ground Ladders,** Ninth Edition. To receive the maximum learning experience from these activities, it is recommended that you use the following procedure:

1. Read the chapter, underlining or highlighting important terms, topics, and subject matter. Study the photographs and illustrations, and read the captions under each.

2. Review the list of vocabulary words to ensure that you know the chapter-related meaning of each. If you are unsure of the meaning of a vocabulary word, look the word up in the glossary or a dictionary, and then study its context in the chapter.

3. On a separate sheet of paper, complete all assigned or selected application and review activities before checking your answers.

4. After you have finished, check your answers against those on the pages referenced in parentheses.

5. Correct any incorrect answers, and review material that was answered incorrectly.

Vocabulary

Be sure that you know the chapter-related meanings of the following words.

- extruded *(24)*
- serrated *(25)*
- knurled *(25)*
- dimpled *(25)*
- corrugated *(25)*
- torsion *(32)*
- toggle *(38)*

Application Of Knowledge

Choose from the following applications those that are appropriate to your department and equipment, or ask your training officer to choose appropriate applications. Practice the chosen applications under your training officer's supervision.

- Inspect wood, metal, or fiberglass ground ladders. *(42, 43)*
- Clean and inspect ground ladders. *(45)*
- Spot refinish a small area of damage to the varnish coat of a wood ground ladder. *(4)*
- Renew the surface coating on a fiberglass ladder. *(45)*
- Perform ground ladder strength tests applicable to all but pompier and folding ladders. *(47)*
- Test the strength of a roof ladder hook. *(49)*
- Service test the hardware on an extension ladder. *(49)*
- Test the strength of a pompier ladder. *(50)*
- Test the strength of a folding ladder. *(51)*

Review Activities

1. List four types of abuse common to fire service ground ladders. *(17)*

2. Identify the following:
   - Coast Douglas fir *(19)*
   - NFPA 1931 *(17)*
   - NFPA 1932 *(17)*
   - tenon *(21)*

- mortise *(21)*
- tongue and groove construction *(23)*
- bushing *(24)*
- A-frame *(30)*
- shim *(38)*

3. Distinguish between solid beam and truss beam ground ladder construction. *(19, 20)*

4. List advantages and disadvantages of truss beam ladder construction. *(20)*

5. List advantages and disadvantages of solid beam ladder construction. *(20)*

6. List advantages and disadvantages of wood ground ladders. *(22, 23)*

7. List advantages and disadvantages of metal ground ladders. *(26, 27)*

8. List advantages and disadvantages of fiberglass ground ladders. *(28)*

9. List advantages and disadvantages of composite wood and metal ground ladders. *(29)*

10. Compare and contrast the four types of combination ladders. *(29-31)*

11. Briefly describe the design and purpose of pompier ladders. *(31, 32)*

12. Compare and contrast the design and operation of enclosed automatic latching and manual latching pawls. *(32-34)*

13. Briefly discuss the design and purpose of the following ladder accessories and hardware:

   - halyard system *(34)*
   - stops *(36)*
   - foot pads *(37)*
   - butt spurs *(36)*
   - tie rod *(37)*
   - toe rod *(37)*
   - staypole system *(38)*
   - mud guard *(38)*
   - protection plate *(38)*
   - levelers *(38)*

14. Identify the contents and typical location of each of the following ladder labels: *(40)*

    - Manufacturer's Identification
    - Serial Number
    - Certification Label
    - Heat Sensor Label
    - Electrical Hazard Warning Label
    - Ladder Positioning Label

15. List the five general items that should be checked when inspecting wood, metal, and fiberglass ground ladders. *(42)*

16. List the areas that should be checked when inspecting wood and wood components of composite ground ladders. *(42)*

17. Describe areas to check when inspecting roof ladders. *(42)*

18. List areas to check when inspecting extension and pole ladders. *(43)*

19. State the actions to take when a ladder is found to be defective: It contains any of the conditions listed or described in Activities 15 - 18. *(43)*

20. Distinguish between the terms *maintenance* and *repair*. *(43)*

21. List the four general maintenance requirements for all types of ladders. *(43, 44)*

22. List maintenance requirements specific to wood ground ladders only. *(44)*

23. List maintenance guidelines specific to extension and pole ladders. *(45)*

24. Explain the differences between design verification testing and service testing of ground ladders. *(47)*

25. List NFPA 1932 requirements pertaining to the frequency of service testing ground ladders. *(47)*

26. Explain why records on ground ladder strength service tests, repairs, and retests are important. *(52)*

Questions And Notes

Chapter 3

Handling Ladders

LEARNING OBJECTIVES

This chapter provides information that will assist the reader in meeting the objectives contained in the Ladders section of NFPA 1001, *Standard for Fire Fighter Professional Qualifications* (1992 edition). The objectives contained in this chapter are as follows:

Fire Fighter I

3-11.2   Carry, position, raise, and lower the following ground ladders:

(a) 14-ft (4.3-m) single or wall ladder

(b) 24-ft (7.3-m) extension ladder

(c) 35-ft (10.7-m) extension ladder

(d) Folding/attic ladder

Handling Ladders

For the purpose of this manual, the term handling ladders refers to the movement of ground ladders from their point of storage on the fire apparatus (or another temporary position) to placement at the point of use. The topics discussed include which ladders are carried on various pieces of apparatus; where they are mounted; and how they are removed, carried, and placed. It is important that these tasks be accomplished in a safe and efficient manner that will not damage either the ladder or other property. Movements need to be smooth and instinctive because speed is essential in many instances. Because more than one firefighter is frequently required, development of teamwork is another important factor. Therefore, proficiency in handling ladders will be realized only with repeated practical training.

LOCATIONS AND METHODS OF MOUNTING ON APPARATUS

While there are standards for the minimum sizes and types of ladders required to be carried on each type of fire apparatus, there are no established standards for the location or the method of mounting the ladders. Locations and methods of mounting vary according to the following factors:

- Manufacturer's policy
- Type of mounting bracket or racking used
- Type of apparatus
- Body design
- Individual fire department requirements
- Type of ladder

These differences make it necessary for each fire department to develop and administer its own training procedures for removing and using ladders.

There are some general rules of thumb that should be observed for the location of ladders on the apparatus. First, and most obvious, ladders should be located in a manner that will facilitate their easy removal from the apparatus. Ladders mounted so high that they cannot be easily reached by an average-sized firefighter are of little use (Figure 3.1). As well, ladders should not be mounted in locations where they will be subjected to engine or exhaust heat. This heat can have a damaging effect on ground ladders, particularly those of the wooden variety.

Figure 3.1 Some ladders are mounted so high on the side of the apparatus that they are difficult to remove. *Courtesy of Joel Woods.*

Ladders Carried On Pumpers

NFPA 1901, *Standard for Pumper Fire Apparatus,* sets the minimum lengths and types of ladders to be carried on all pumper or engine companies. Each pumper must carry the following ladders:

- One 10-foot (3 m) folding ladder
- One 14-foot (4.3 m) roof ladder
- One 24-foot (8 m) or larger extension ladder

Many fire departments prefer to carry a 35-foot (11 m) three-section extension ladder instead of the 24-foot (8 m) ladder. Regardless of which size is selected, ladders on pumpers are usually mounted vertically (on the beam) on the right (passenger) side of the apparatus (Figure 3.2). This racking location usually results in one end of the ladder overhanging the rear of the apparatus. A padded guard should be placed over the protruding ends to prevent injury to persons should they accidentally walk into the end of the ladder (Figure 3.3).

Figure 3.2 Most pumpers carry their ladders mounted on the right side of the apparatus.

Figure 3.3 Ladder pads reduce the potential for injuries should someone walk into the back of the ladder.

When a pumper is constructed with high side compartments on both sides of the apparatus, ladders can be mounted in an overhead rack (Figure 3.4). An overhead rack is also used when it is necessary to carry a ladder longer than 35 feet (11 m) on a pumper. Other pumpers have a rack atop or above the side compartments (Figure 3.5). Some of these are on electrically or hydraulically driven arms that swing the ladders down to a convenient height for removal (Figures 3.6 a through c).

Figure 3.4 Some apparatus have ladders mounted on fixed racks above the hose bed.

Figure 3.5 Overhead racks may also be directly above the side compartments.

Figure 3.6a A hydraulic ladder rack in the stowed position.

Figure 3.6b The hydraulic ladder rack slowly swings into the stowed position.

Figure 3.6c The ladders become vertical when the hydraulic ladder rack is completely lowered.

Ladders Carried On Aerial Apparatus

Because the placement of ground ladders is one of the primary functions of a ladder or truck company, aerial apparatus carry a larger complement of ground ladders than pumpers carry. NFPA 1904, *Standard for Aerial Ladder and Elevating Platform Fire Apparatus* (1991) specifies that, as a minimum, the following ground ladders must be carried on aerial apparatus:

- One 10-foot (3 m) folding ladder
- Two 16-foot (5 m) roof ladders
- One 14-foot (4.3 m) combination ladder
- One 24-foot (8 m) extension ladder
- One 35-foot (11 m) extension ladder

The 1991 edition of NFPA 1904 greatly reduced the minimum number of ground ladders required to be carried on aerial apparatus, when compared to previous editions of the standard (which was formerly numbered NFPA 1901 and contained the requirements for all types of apparatus; the 1991 edition of NFPA 1901 contains only requirements for pumpers). The old edition required that several additional extension ladders be carried, including at least one 40-foot (12 m) or larger pole ladder. Because many aerial apparatus still in service today were purchased during the tenure of the previous edition of the standard, many will have more ground ladders than those listed above. As well, many fire departments became accustomed to having the larger number of ground ladders available and still specify more than the minimum when acquiring new apparatus.

The racking arrangement of ground ladders on aerial apparatus is frequently influenced by the amount of space needed for the mounting of the aerial device. Because of this space requirement, a variety of racking schemes are used to carry the required number, types, and lengths of ground ladders. The following list describes some of these arrangements:

- *Loaded from the rear, lying flat in tiers usually two rows wide and two rows high* (Figure 3.7). This type racking requires the ladders to fit into runners or troughs. The ladders are locked into position at each tier by either a manually operated lever or a power-operated bar activated by an electric solenoid. When ladders are mounted in this manner, room must be left at the rear of the apparatus to allow for removal and loading.

- *Nested vertically (on edge) or in flat tiers on the sides of the apparatus* (Figures 3.8 a and b). This arrangement eliminates the problem of another piece of apparatus stopping too close and preventing the ladder from being removed.

- *Nested vertically (on edge) on each side (Figure 3.9).* These ladders are arranged so that they may be loaded and unloaded from the rear of the apparatus by sliding them in and out of the troughs. When ladders are mounted in this manner, room must be left at the rear of the apparatus to allow for removal and loading.

Figure 3.7 These ladders are loaded in a flat position from the rear of the apparatus.

Figure 3.8a An example of an aerial apparatus with ladders mounted vertically on the side of the apparatus.

Figure 3.8b Some aerial apparatus have ladders stacked flat on the side of the apparatus.

Figure 3.9 Some ladders that are mounted vertically on the side of the apparatus must be loaded/unloaded from the rear.

Folding ladders, because of their compactness, are mounted wherever it is convenient (Figure 3.10). Combination ladders, depending on their exact design, may be located on the side of the apparatus, in a topside cargo hold, or in a large compartment.

Figure 3.10 A folding ladder in a typical mounting position on the apparatus.

Ladders Carried On Other Apparatus

Ground ladders are not limited to pumpers and aerial apparatus. They are commonly carried on a variety of other fire department vehicles. The following sections briefly highlight some of these vehicles.

INITIAL ATTACK APPARATUS

NFPA 1902, *Standard for Initial Attack Fire Apparatus,* contains requirements for vehicles that are more commonly referred to as mini- or midi-pumpers. These are small, lightweight apparatus intended to respond to and control a fire prior to the arrival of larger apparatus (Figure 3.11). Because of their small size, they are not required to carry the same complement of equipment required of full-sized pumpers. NFPA 1902 sets the following requirements for ground ladders on initial attack apparatus:

Figure 3.11 Minipumpers may carry small ground ladders. *Courtesy of Joel Woods.*

- Vehicles with a gross vehicle weight (GVW) less than 15,000 pounds (6 810 kg): One 12-foot (4 m) ladder of any type

- Vehicles with a GVW between 15,000 and 20,000 pounds (6 810 kg and 9 080 kg): One 14-foot (4.3 m) ladder of any type

- Vehicles with a GVW in excess of 20,000 pounds (9 080 kg): One 16-foot (5 m) ladder of any type

MOBILE WATER SUPPLY APPARATUS

NFPA 1903, *Standard for Mobile Water Supply Fire Apparatus,* contains the requirements for these vehicles, which are more commonly known as tankers or tenders. Tankers (tenders) that are designed solely to transport water are not required to carry ground ladders (Figure 3.12). Some tankers (tenders) are equipped with a fire pump and are designed to operate in much the same way as a standard pumper or engine company. These vehicles are commonly called pumper/tankers or pumper/tenders. As a minimum, these vehicles should be equipped with the same ground ladders specified for pumpers in NFPA 1901 (Figure 3.13).

Figure 3.12 Elliptical tankers usually do not carry ground ladders. *Courtesy of Joel Woods.*

Figure 3.13 Pumper/tankers frequently carry ground ladders. *Courtesy of Joel Woods.*

RESCUE OR SQUAD APPARATUS

Rescue or squad apparatus may, in some cases, be equipped with ground ladders to facilitate their operations in the event another fire apparatus is not on the scene with them (Figure 3.14). At the time this manual was published, no NFPA standard existed for the design of this type of apparatus. The decision to carry ladders on this type of apparatus, the types and sizes of ladders to be carried, and where they will be carried on the apparatus rest solely on the preferences of each particular fire department.

Figure 3.14 Large rescue vehicles may carry ground ladders. *Courtesy of Joel Woods.*

BASIC LADDER HANDLING TECHNIQUES

In order to be successful, firefighters must be able to estimate what ladder will be required to do the given job, select the proper ladder from those available, and quickly remove the ladder from the apparatus. The following sections provide information that will assist the firefighter in accomplishing these tasks.

Selecting The Correct Ladder For The Job

Selecting a ladder to do a specific job requires that the firefighter be a good judge of distance. Roughly speaking, a residential story will average 8 to 10 feet (2.5 m to 3 m), and the distance from the floor to the windowsill about 3 feet (1 m). A commercial story will average 12 feet (4 m) from floor to floor, and the distance from the floor to the windowsill about 4 feet (1.2 m). In general, Table 3.1 can be used in selecting ladders.

TABLE 3.1 Ladder Selection Guide	
First story roof	16 to 20 feet (4.9 m to 6.0 m)
Second story window	20 to 28 feet (6.0 m to 8.5 m)
Second story roof	28 to 35 feet (8.5 m to 10.7 m)
Third story window or roof	40 to 50 feet (12.2 m to 15.2 m)
Fourth story roof	over 50 feet (15.2 m)

Working rules for ladder length include the following:

- The ladder should extend a few feet (preferably five rungs) beyond the roof edge to provide both a footing and a handhold for persons stepping on or off the ladder (Figure 3.15).

- When used for access from the side of a window or for ventilation, the tip of the ladder should be placed three or four rungs above the windowsill (Figure 3.16).

- When rescue from a window opening is to be performed, the tip of the ladder should be placed just below the windowsill (Figure 3.17).

The next step is to determine how far various ladders will reach. Knowledge of the designated length of a ladder can be used to answer this question. Remember that the designated length (this figure is normally displayed on the ladder) is derived from a measurement of the maximum extended length (Figure 3.18). This is NOT THE LADDER'S REACH because ladders are set at angles of approximately 75 degrees. Reach will therefore be LESS than the designated length. One more thing needs to be considered: Single, roof, and folding ladders meeting NFPA 1931 are required to have a measured length equal to the designated length. However, the maximum extended length of extension and pole ladders may be as much as 6 inches (150 mm) LESS than the designated length.

Figure 3.16 A ladder placed for window ventilation.

Figure 3.17 A ladder used for a window rescue should have the tip placed just below the sill.

Figure 3.18 The ladder length is marked on the outside of both beams at the butt.

Table 3.2 provides information on the reach of various ground ladders when placed at the proper climbing angle. However, the following should be noted when considering the information contained in Table 3.2:

- For lengths of 35 feet (11 m) or less, reach is approximately 1 foot (300 mm) less than the designated length.

- For lengths over 35 feet (11 m), reach is approximately 2 feet (600 mm) less than the designated length.

Figure 3.15 A ladder raised to a roof should have at least five rungs above the roof level.

TABLE 3.2
Maximum Working Heights For Ladders Set At Proper Climbing Angle

Designated Length of Ladder		Maximum Reach	
10 foot	(3.0 m)*	9 feet	(2.7 m)*
14 foot	(4.3 m)	13 feet	(4.0 m)
16 foot	(4.9 m)	15 feet	(4.6 m)
20 foot	(6.1 m)	19 feet	(5.8 m)
24 foot	(7.3 m)	23 feet	(7.0 m)
28 foot	(8.5 m)	27 feet	(8.2 m)
35 foot	(10.6 m)	34 feet	(10.4 m)
40 foot	(12.2 m)	38 feet	(11.6 m)
45 foot	(13.7 m)	43 feet	(13.1 m)
50 foot	(15.2 m)	48 feet	(14.6 m)

RULE OF THE THUMB

Ladders 35 feet (11 m) and under reach 1 foot (.3 m) less than the designated lengths.

Ladders over 35 feet (11 m) reach 2 feet (.6 m) less than the designated length.

Elsewhere in this text metrics have been rounded off to the nearest whole number. In this instance metrics are rounded off to the nearest tenth to more accurately show the difference.

For purposes of figuring ladder length needed, the differences between the designated length and reach will be known as the *reach factor*. Some examples will help in understanding the practical application of this information.

Example 1: You are directed to ladder the eaves of a two-story residence. Assuming that each story is 8 feet (2.5 m), what ladder length will be required?

Feet (Meters) per story	8 feet	(2.5 m)
Number of stories	x 2	x 2
	16 feet	(5 m)
Length needed for handhold	4 feet	(1.2 m)
Reach Factor	1 foot	(0.3 m)
Length of ladder required	21 feet	(6.5 m)

Realistically, a 24-foot (8 m) extension ladder will be needed.

Example 2: You are directed to place a ladder for access into a third floor window at a boarding house. Assuming that each story is 10 feet (3 m), what ladder length will be required?

Feet (Meters) per story	10 feet	(3 m)
Number of stories	x 2	x 2
	20 feet	(6 m)
Length needed for handhold	4 feet	(1.2 m)
Floor-to-windowsill height for 3rd floor	3 feet	(1 m)
Reach Factor	1 foot	(0.3 m)
Length of ladder required	28 feet	(8.5 m)

Although a 28-foot (8.5 m) extension ladder may just fit, it would probably be best to use a 35-foot extension ladder for this situation.

Example 3: What length of ladder will be required to perform a window rescue from a fourth floor window in a commercial occupancy?

Feet (Meters) per story	12 feet	(4 m)
Number of stories	x 3	x 3
	36 feet	(12 m)
Floor-to-windowsill height for 4th floor	4 feet	(1.2 m)
Reach Factor	2 feet	(0.6 m)
Length of ladder required	42 feet	(13.8 m)

In this case, a 45-foot (13.5 m) pole ladder will be required to perform the rescue.

When all factors are considered, the 35-foot (11 m) extension ladder emerges as the most versatile of all extension ladders. This is because it is suitable for one- to three-story residential, one- and two-story commercial, and one-story industrial structures; and it is compact enough to be carried on the side of the pumper.

Ladders carried on aerial apparatus are sometimes preferred by firefighters because they are frequently wider than the ladders carried on pumpers, making them easier to work from. As well, aerial apparatus may carry two-section 35-foot (11 m) extension ladders, as opposed to the three-section ladders carried on pumpers. The two-section ladder from the aerial apparatus will be lighter than the three-section ladder from the pumper.

Removing Ladders From Apparatus

Ladders are secured on the apparatus by various means, and the method of releasing the locking mechanism is unique to each manufacturer (Figures 3.19 a and b). It is necessary for each firefighter to determine how the releasing mechanisms on the department apparatus operate. Once the securing device is released, there are other considerations to take into account:

Figure 3.19a Most ladder-holding brackets are spring loaded.

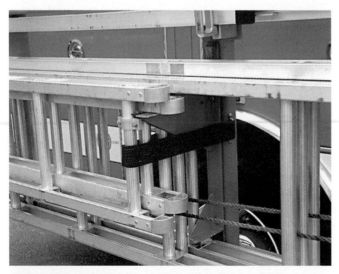

Figure 3.19b Some newer apparatus may utilize Velcro® straps to restrain ladders.

- Ladders are usually carried nested, with more than one in a racking, so it may be necessary to remove one or more to get to the ladder needed.

- Removal of any ladder from the racking frequently leaves the securing device nonfunctional. The remaining ladder(s) may fall from the apparatus, particularly if the vehicle is moved or if it is vibrating. Apparatus with ladders mounted vertically on the side are prone to this problem. Some have a separate holding device for each ladder to prevent this from happening.

- Unused ladders should be put back on the apparatus and secured or stored in a safe and available location where individuals will not trip over them, where vehicles will not run over them, and where they are away from engine and exhaust heat (Figure 3.20). Departments should adopt a standard policy governing this procedure so that all personnel will perform in a like manner.

Figure 3.20 Unused ladders should be placed back on the brackets.

Before firefighters are drilled in removing ground ladders from apparatus, each firefighter should be able to answer the following questions:

- What ladders (types and lengths) are carried and where are they carried on the apparatus?

- Are the ladders racked with the butt toward the front or toward the rear of the apparatus?

- Where ladders are nested together, can one ladder be removed leaving the other(s) secured in place? (In particular can the roof ladder be removed from the side of the pumper and leave the extension ladder secured in place?)

- In what order do the ladders that nest together rack? (Pumper extension ladder goes on first, roof ladder second, or vice versa?)

- Is the top fly of the extension ladder on the inside or on the outside when the ladder is racked on the side of the apparatus?

- How are the ladders secured?

- When ladders are mounted vertically on the side of apparatus, which rungs go in or near the brackets? (Many departments find it a good practice to mark ladders to indicate when rungs go in or near the brackets, as shown in Figure 3.21.)

Figure 3.21 Most ladders have marks on them to denote where they should be placed on the mounting brackets.

Firefighters should not attempt to remove ladders from any apparatus equipped with an aerial device while the stabilizers are being deployed. This practice prevents the firefighter from being struck by a moving stabilizer. If the aerial device is in a deployed position, make sure that it is not in contact with any charged electrical lines or equipment before attempting to remove the ladder.

> # WARNING
> **Attempting to remove a ladder from, or even touching, an aerial apparatus that is in contact with live electrical equipment may result in electrocution of the firefighter(s) coming in contact with the apparatus.**

Proper Lifting And Lowering Methods

Each year many firefighters are injured when using improper lifting and lowering techniques. Often, these injuries are preventable. The following procedures are recommended:

- Have adequate personnel for the task.

- Bend knees, keeping back as straight as possible, and lift with the legs NOT WITH THE BACK OR ARMS (Figure 3.22).

- When two or more firefighters are lifting a ladder, lifting should be done on the command of a firefighter at the rear who can see the whole operation (Figure 3.23). If any firefighter is not ready, that person should make it known immediately so that the operation will be halted. Lifting should be done in unison.

Figure 3.22 Always lift with your knees, not your back.

Figure 3.23 The firefighter at the butt gives the command to lift.

- When it is necessary to place a ladder on the ground before raising it, the reverse of the procedure for lifting is used. Lower the ladder with the leg muscles. Also, be sure to keep the body and toes parallel to the ladder so that when the ladder is placed it does not injure the toes (Figure 3.24).

The procedures for initiating ladder carries for ladders lying on the ground differ from those for ladders that are carried on the apparatus. Different storage methods require different procedures that must be adapted to the individual situation.

NOTE: Whenever the following text describes a carry initiated from the racking on apparatus, it presumes that the securing device has been released.

Figure 3.24 Keep your body and toes parallel to the ladder as it is lowered.

ONE-FIREFIGHTER CARRIES

Single or roof ladders may be safely carried by any single firefighter. Many firefighters are also capable of safely carrying a 24-foot (8 m) extension ladder single-handedly; however, it is more desirable to use two firefighters on any extension ladder.

There are three methods by which one firefighter may carry a ladder: the low-shoulder method, the high-shoulder method, and the arm's length method. Except where noted, the ladder butt is always carried forward.

Low-Shoulder Method

The low-shoulder carry involves resting the ladder's upper beam on the firefighter's shoulder, while the firefighter's arm goes between two rungs (Figure 3.25).

Figure 3.25 A completed one-firefighter low-shoulder carry.

WARNING

The forward end of the ladder is carried slightly lowered. Lowering the forward portion provides better balance when carrying, improves visibility by allowing the firefighter to view the way ahead, and if the ladder should strike another person, the butt spurs will make contact with the body area instead of the head.

The following sections show the steps for performing the low-shoulder carry from various racking locations and from flat on the ground.

FROM VERTICAL RACKING/SIDE REMOVAL

Step 1: Face the side of the apparatus, near the center (lengthwise) of the ladder. This position will be on the balance point for the carry. (NOTE: Some fire departments mark this point on the ladder for convenience in initiating one-firefighter carries.)

Step 2: Grasp the most convenient rungs of the ladder (Figure 3.26).

Step 3: Lift the ladder free of the rack, and at the same time, take a step backward (Figure 3.27).

Step 4: Release the rung toward the tip. Pivot toward the butt, simultaneously inserting your free arm between the rungs, and bring the upper beam onto your shoulder (Figure 3.28). Grasp two convenient rungs, one with each hand, and begin to walk.

FROM VERTICAL RACKING/REAR REMOVAL

Step 1: Grasp the bottom rung and pull the ladder straight back, halting when the ladder is just far enough in the rack that it will remain there without your support (Figure 3.29).

Figure 3.28 Simultaneously pivot toward the butt and insert your free arm between two rungs.

Figure 3.26 Grasp two rungs.

Figure 3.29 Pull the ladder about one-half way out of the rack.

Step 2: Release your grip on the rung and proceed to the ladder midpoint. Insert your near arm between the rungs, grasp a convenient rung, and proceed forward with the ladder. When the ladder clears the rack sufficiently, it is settled upon your shoulder (Figure 3.30).

Figure 3.27 Lift the ladder free of the rack and step back.

Figure 3.30 Insert an arm between two rungs and lift the upper beam onto your shoulder.

FROM FLAT RACKING/SIDE REMOVAL

Step 1: Take a position alongside the apparatus, facing the midpoint of the ladder. Grasp the ladder by the near beam, and pull it outward.

Step 2: Before the ladder clears the rack, shift your grip to two convenient rungs (Figure 3.31).

Step 3: Pull the ladder clear of the rack, and allow the far beam to swing downward until the ladder is vertical (Figure 3.32).

Step 4: Release the rung toward the tip. Pivot toward the butt, simultaneously inserting your free arm between the rungs, and bring the upper beam onto your shoulder.

Figure 3.31 Shift your grip to two convenient rungs before the ladder clears the rack.

Figure 3.32 Allow the ladder to swing down to a vertical position.

FROM FLAT RACKING/REAR REMOVAL

Step 1: Pull the ladder nearly out of the rack and lower the butt to the ground. (**NOTE:** Ladders racked overhead may require lowering the tip from the edge of the rack to the tailboard or to an intermediate step.)

Step 2: Take a position at midpoint, facing the ladder. Grasp two convenient rungs (Figure 3.33).

Step 3: Bring the far beam upward as your body is pivoted toward the butt. Insert the arm nearest the tip end between two rungs, and bring the ladder onto your shoulder.

Figure 3.33 Grasp two rungs.

FROM FLAT ON THE GROUND

Step 1: Kneel beside the ladder, facing the tip. Grasp the middle rung with your near hand, palm facing forward (Figure 3.34).

Step 2: Lift the ladder. As the ladder rises, pivot into the ladder, placing your free arm between two rungs so that the upper beam comes to rest on your shoulder (Figure 3.35).

Figure 3.34 Kneel facing the tip and grasp a rung with your palm facing forward.

Figure 3.35 Simultaneously lift the ladder, pivot toward the butt, and insert your free arm between two rungs.

High-Shoulder Method

The one-firefighter high-shoulder carry is particularly well suited for making beam raises (Figure 3.36). It is not used for carrying a roof ladder when it is to be taken up another ladder.

Figure 3.38 Lift and pivot toward the butt, placing the lower beam onto your shoulder.

Figure 3.36 A completed high-shoulder carry.

Figure 3.39 Shift the bottom hand to the outside of the lower beam.

FROM VERTICAL RACKING/SIDE REMOVAL

Step 1: Face the ladder at its midpoint. Use your hand nearest the butt to grip the top beam. Lift the ladder slightly so that your other hand can be placed with the palm up under the bottom beam (Figure 3.37).

Step 2: Lift the ladder free of the rack. Simultaneously pivot toward the butt and bring the bottom beam to rest on your shoulder (Figure 3.38).

Step 3: Shift the hand holding the bottom beam to grasp the outside of that beam (Figure 3.39).

FROM VERTICAL RACKING/REAR REMOVAL

Step 1: Pull the ladder most of the way out of the rack. Position yourself at the midpoint of the ladder (Figure 3.40).

Step 2: Lift the ladder, pivoting toward the butt, and place the bottom beam on your shoulder (Figure 3.41).

Step 3: Shift your handholds so that one hand is grasping the top beam and the other either the bottom beam or a convenient rung at a point next to the bottom beam.

Figure 3.37 The hand nearest the butt should be on the top beam, and the hand nearest the tip should be on the lower beam.

Figure 3.40 Pull the ladder out to its midpoint.

Figure 3.41 Place the lower beam on your shoulder.

FROM FLAT RACKING/SIDE REMOVAL

Step 1: Take a position alongside the apparatus, facing the midpoint of the ladder. Grasp the ladder by the near beam, and pull it outward.

Step 2: As the ladder reaches the edge of the rack, tilt the outside beam downward onto your shoulder (Figure 3.42).

Step 3: Tilt the ladder up until it is vertical, and adjust so that your near hand is grasping the outside of the lower beam and your far hand is grasping the top beam or high up on a rung (Figure 3.43).

Figure 3.42 Tilt the outside beam onto your shoulder.

Figure 3.43 Pull the ladder completely onto your shoulder.

FROM FLAT RACKING/REAR REMOVAL

Step 1: Pull the ladder nearly out of the rack, and lower the butt to the ground (Figure 3.44).

Step 2: Take a position at midpoint, facing the ladder. Grasp a convenient rung and the near beam (Figure 3.45).

Step 3: Lift the ladder and simultaneously turn it to vertical while pivoting under the bottom beam and placing it on your shoulder (Figure 3.46).

FROM FLAT ON THE GROUND

Step 1: Kneel beside the ladder, facing the tip. Grasp the middle rung with your near hand, palm facing forward (Figure 3.47).

Step 2: Lift the ladder, simultaneously turning it to vertical and pivoting under the bottom beam, which is placed on your shoulder (Figure 3.48).

Figure 3.44 Pull the ladder out and lay the butt on the ground.

Figure 3.45 Grasp a rung with the hand nearest the butt and grasp the beam with the hand nearest the tip.

Figure 3.46 Bring the ladder up to shoulder level.

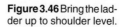

Figure 3.47 Kneel facing the tip.

Figure 3.48 Lift, pivot toward the butt, and place the lower beam onto your shoulder.

Figure 3.50 Set one beam of the ladder on the ground.

Figure 3.51 Pick up the top beam.

Arm's Length Method

The one-firefighter arm's length carry is best suited for a lighter weight ladder because all of the ladder's weight is carried in one hand (Figure 3.49). Control of the ladder's side-to-side movement is minimal with this carry and there is a tendency for the ladder to bump against the leg.

Figure 3.49 A completed arm's length carry.

FROM VERTICAL RACKING/SIDE REMOVAL

Step 1: Face the ladder at midpoint. Grasp two convenient rungs. Lift the ladder out of the rack, and place it on the ground, resting it on one beam (Figure 3.50).

Step 2: Pivot toward the butt, crouch slightly, and grasp the upper beam with one hand (Figure 3.51).

Step 3: Lift the ladder off the ground, using the leg muscles.

FROM VERTICAL RACKING/REAR REMOVAL

Step 1: Pull the ladder most of the way out of the rack. Position yourself at the midpoint of the ladder.

Step 2: Grasp two convenient rungs. Pull the ladder clear of the rack, and place it on the ground in the vertical position (Figure 3.52).

Step 3: Pivot toward the butt. Grasp the upper beam with one hand (Figure 3.53).

Step 4: Lift the ladder off the ground, using the leg muscles.

Figure 3.52 Set one beam of the ladder on the ground.

Figure 3.53 Pick up the top beam.

FROM FLAT RACKING/SIDE REMOVAL

Step 1: Take a position alongside the apparatus, facing the midpoint of the ladder. Grasp the ladder by the near beam, and pull it outward.

Step 2: Before the ladder clears the rack, shift your hand nearest the butt to grasp a rung (Figure 3.54).

Step 3: Pull the ladder clear of the rack and bring it downward, simultaneously tilting it to vertical (Figure 3.55).

Step 4: Release your grip on the rung and pivot toward the butt.

Figure 3.54 Pull the ladder out of the rack.

Figure 3.55 Tilt the ladder down until it is vertical.

FROM FLAT RACKING/REAR REMOVAL

Step 1: Pull the ladder from the rack (Figure 3.56).

Step 2: When the ladder clears the rack, lower it and bring it to vertical (Figure 3.57).

Step 3: With the hand nearest the butt, release your grip and pivot toward the butt.

Figure 3.56 Pull the ladder from the rack.

Figure 3.57 Lower the ladder and bring it to vertical.

FROM FLAT ON THE GROUND

Step 1: Crouch, facing the ladder at midpoint. Tilt the ladder up onto one beam (Figure 3.58).

Step 2: Grasp the upper beam and stand up, at the same time pivoting toward the butt (Figure 3.59). The hand nearest the butt is released to complete the evolution.

Figure 3.58 Tilt the ladder onto one beam.

Figure 3.59 Stand up and face the butt.

Special Procedures For Carrying Roof Ladders

Procedures previously described are for ladders that are carried butt forward. In some cases, a firefighter will carry a roof ladder with the intention of climbing another ground ladder and placing

the roof ladder with hooks deployed on a sloped roof. In this situation, the firefighter should use the low-shoulder method and have the tip (hooks) forward (Figure 3.60).

When the tip is carried forward, *the pivot is made toward the tip*. When the ladder is to be picked up from the ground, the firefighter will have to kneel facing the butt instead of the tip (Figure 3.61).

Normally, the roof ladder is carried, with the hooks closed, to the foot of the second ladder. A second firefighter opens the hooks while the first firefighter maintains the carry (Figure 3.62). When no second firefighter is present, the firefighter sets

Figure 3.60 The roof ladder may be carried with the tip forward.

Figure 3.61 Kneel facing the butt.

Figure 3.62 A second firefighter may open the roof hooks.

the ladder down, moves to the tip, picks up the tip, opens the hooks, lays the tip down, returns to the midpoint, picks up the ladder, and resumes the carry (Figures 3.63 a and b).

There may be occasions when there is no second firefighter to open the hooks, time is critical, and there is no crowd of people through which the ladder must be carried. In this case, the hooks may be opened at the apparatus before the carry is begun; they are turned outward in relation to the firefighter carrying the ladder (Figures 3.64 a and b).

Figure 3.63a Open the hooks.

Figure 3.63b Return to the carrying position.

Figure 3.64a Turn the hooks out while the ladder is still near the apparatus.

Figure 3.64b Carry the ladder with the hooks open facing outward.

TWO-FIREFIGHTER CARRIES

Although they may be used with single or roof ladders, two-firefighter carries are most commonly used for 24-, 28- and 35-foot (8 m, 8.5 m, and 11 m) extension ladders. The three methods most commonly used when two firefighters carry a ladder are the low-shoulder method, the hip or underarm method, and the arm's length on-edge method.

Low-Shoulder Method

The two-firefighter low-shoulder carry gives firefighters excellent control of the ladder (Figure 3.65). The forward firefighter places his or her free hand over the upper butt spur. This is done to prevent injury in case there is a collision with someone while the ladder is being carried. This measure can be taken with most of the two-firefighter carries.

Figure 3.65 A completed two-firefighter low-shoulder carry.

FROM VERTICAL RACKING/SIDE REMOVAL

Step 1: Both firefighters stand facing the ladder, one near the tip and the other near the butt. Each firefighter uses both hands to grasp the ladder and remove it from the rack (Figure 3.66).

Step 2: The firefighters continue to grasp the ladder with their hands nearest the butt while each pivots, places their other arm between two rungs, and brings the upper beam onto their shoulders (Figure 3.67).

Figure 3.66 Face the ladder and grasp the last two rungs on either end.

Figure 3.67 Pivot toward the butt while inserting your free arm between the last two rungs.

FROM VERTICAL RACKING/REAR REMOVAL

Step 1: One firefighter grasps the bottom (end) rung and pulls the ladder from the rack. The second firefighter stands to the side of the ladder, adjacent to the rear of the apparatus, and assists with removal (Figure 3.68).

Step 2: When the ladder is almost out of the rack, the first firefighter shifts to a position adjacent to the butt on the same side as the second firefighter (Figure 3.69).

Step 3: Both firefighters turn to face the butt, simultaneously placing the arm on the side toward the apparatus between the two bottom and the two top rungs, respectively (Figure 3.70). The firefighters pull the ladder clear of the rack and bring it onto their shoulders.

Figure 3.68 One firefighter pulls the ladder from the rack.

Figure 3.69 The first firefighter shifts to the same side of the ladder as the second firefighter.

Figure 3.70 Face the butt and insert each arm between two rungs.

FROM FLAT RACKING/SIDE REMOVAL

Step 1: The firefighters position themselves facing the ladder, one at each end. Both grasp the beam and pull the ladder outward. As soon as the ladder is part way out, the firefighters shift at least one hand to grip a rung (Figure 3.71).

Step 2: The firefighters pull the ladder clear of the rack and bring it toward vertical by swinging the far beam downward toward them. Both firefighters pivot toward the butt, simultaneously placing the arm on the side toward the tip between two rungs and bring the ladder onto their shoulders (Figure 3.72).

FROM FLAT RACKING/REAR REMOVAL

Step 1: The two firefighters pull the ladder from the rack far enough so that they can position themselves, one at each end, on the same side, facing the ladder. Each firefighter grasps two rungs of the ladder (Figure 3.73). The height of the ladder as it is pulled from the apparatus will determine whether the ladder is grasped from above or from beneath the ladder.

Step 2: The firefighters pull the ladder clear of the rack. If the ladder was grasped from underneath, the outside beam is lowered as the inside beam is raised to their shoulders (Figure 3.74). This process is reversed if the ladder was grasped from the top. At the same time, both firefighters pivot toward the butt and place the near arm between two rungs.

FROM FLAT ON THE GROUND

Step 1: The two firefighters kneel on the same side of the ladder, one near the butt and the other near the tip, both facing the tip.

Step 2: The two firefighters grasp a convenient rung with their near hands, palms forward, and set the ladder up on edge (Figures 3.75 a and b).

Step 3: The firefighter at the butt gives the command to "shoulder the ladder." Both firefighters stand up, using their leg muscles to lift the ladder.

Figure 3.71 Grasp the rung and pull the ladder out.

Figure 3.72 As the ladder swings to a vertical position, each firefighter inserts his or her free arm between two rungs.

Figure 3.73 Each firefighter grasps the two end rungs on his or her end of the ladder.

Figure 3.74 Bring the ladder to shoulder level.

Figure 3.75a Begin with the ladder flat on the ground, facing the tip.

Figure 3.75b Stand the ladder on one beam.

Step 4: As the ladder and the firefighters rise, the far beam is tilted upward. Both firefighters pivot and place their free arms between two rungs (Figure 3.76). The upper beam is placed on their shoulders with the firefighters facing the butt. (NOTE: The lift should be smooth and continuous.)

Figure 3.76 The two firefighters stand, pivot, and insert their arms between the end rungs.

Hip Or Underarm Method

The two-firefighter hip or underarm carry is used on single or two-section ladders (Figure 3.77). This carry is not suitable for three-section ladders because of their bulk.

Figure 3.77 A completed two-firefighter hip carry.

FROM VERTICAL RACKING/SIDE REMOVAL

Step 1: The firefighters position themselves near the tip and butt, facing the ladder.

Step 2: The firefighters grasp the ladder with both hands and lift it off the rack. It is then lowered until the top beam is approximately chest high (Figure 3.78).

Step 3: Each firefighter releases the hand closest to the tip end, reaches across the top beam while pivoting the body toward the butt, and grasps a rung (Figure 3.79).

Step 4: Each firefighter places the top beam against his or her body just under the armpit. The firefighters should now be facing the butt.

Figure 3.78 Lower the ladder to chest level.

Figure 3.79 The firefighters let go of the ladder with their hands closest to the tip, reach over the top beam, and grasp a rung.

FROM VERTICAL RACKING/REAR REMOVAL

Step 1: One firefighter grasps the bottom rung and pulls the ladder from the rack. The second firefighter stands to the side of the ladder, adjacent to the back of the apparatus, and assists with the removal (Figure 3.80).

Step 2: When the ladder is almost out of the rack, the first firefighter shifts to a position adjacent to the butt on the same side as the second firefighter (Figure 3.81).

Figure 3.80 One firefighter pulls the ladder from the rack.

Figure 3.81 The first firefighter shifts to the same side of the ladder as the second firefighter.

Step 3: The firefighters grasp the rungs with both hands and lift the ladder clear of the rack. The ladder is then lowered until the top beam is approximately chest high (Figure 3.82).

Figure 3.82 Grasp the ladder by the rungs and bring it to chest level.

Step 4: Each firefighter releases the hand farthest from the butt, reaches across the top beam while pivoting the body toward the butt, and grasps a rung.

Step 5: Each firefighter places the top beam against his or her body just under the armpit. The firefighters should now be facing the butt.

FROM FLAT RACKING/SIDE REMOVAL

Step 1: Firefighters position themselves facing the ladder, one at each end. Both grasp the beam and pull the ladder outward. As soon as the ladder is part way out, the hands at the butt are shifted to grip a rung (Figure 3.83).

Step 2: The firefighters pull the ladder clear of the rack and bring it toward vertical by swinging the far beam downward toward them (Figure 3.84).

Step 3: The firefighters lower the ladder until the top beam is approximately chest high.

Figure 3.83 Pull the ladder out by the bottom rung.

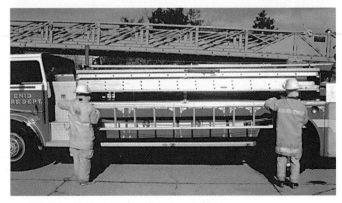

Figure 3.84 Bring the ladder to a vertical position at chest level.

Step 4: Each firefighter releases the hand farthest from the butt, reaches across the top beam while pivoting the body toward the butt, and grasps a rung.

Step 5: Each firefighter places the top beam against his or her body just under the armpit. The firefighters should now be facing the butt.

FROM FLAT RACKING/REAR REMOVAL

Step 1: After pulling the ladder almost from the rack, both firefighters position themselves on the same side of the ladder.

Step 2: The firefighters continue pulling the ladder until it clears the rack. As the ladder clears the rack, each firefighter uses the hand closest to the tip to grasp a rung near to the closest beam.

Step 3: The hand of each firefighter nearest the butt reaches under the ladder and grasps either the rung close to the far beam or the far beam itself (Figure 3.85).

Step 4: The firefighters pivot toward the butt, simultaneously allowing the far beam to swing downward while the near beam swings up against the body just under the armpit (Figure 3.86).

FROM FLAT ON THE GROUND

Step 1: Both firefighters position themselves on the same side of the ladder, facing the butt. One is positioned near the tip end, the other near the butt.

Step 2: The firefighters kneel beside the ladder and each reaches across and grasps a rung near the far beam (Figure 3.87).

Step 3: The firefighters tilt up the ladder so that it is resting on what was the near beam (Figure 3.88).

Step 4: Each firefighter then reaches across the upper beam and grasps a convenient rung (Figure 3.89).

Step 5: The firefighters stand and simultaneously lift the ladder against the body with the upper beam located just under the armpit. (CAUTION: It is important to lift with the leg muscles, keeping the back straight.)

Figure 3.87 Kneel facing the butt.

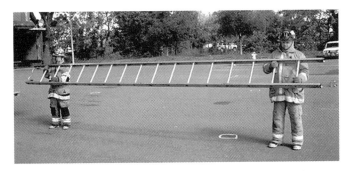

Figure 3.85 Prepare to bring the ladder to vertical at chest level.

Figure 3.88 Stand the ladder on one beam.

Figure 3.86 Pivot and bring the near beam up underneath your armpit.

Figure 3.89 Reach across the top beam and grasp a rung.

Arm's Length On-Edge Method

The two-firefighter arm's length on-edge carry is best performed with lightweight ladders (Figure 3.90). All of the following evolutions are based on the fact that the firefighters are positioned on the bed section (widest) side of the ladder when it is in the vertical position.

Figure 3.90 A completed two-firefighter arm's length carry.

FROM VERTICAL RACKING/SIDE REMOVAL

Step 1: The two firefighters stand facing the ladder, one at each end. The firefighters grasp the rungs with both hands, and lift the ladder from the rack.

Step 2: The two firefighters, without bending over, lower the ladder as far as possible (Figure 3.91).

Figure 3.91 The firefighters lower the ladder to waist level without bending.

Step 3: While retaining the grasp on a rung with the hand nearest the butt, each firefighter releases the hold with the other hand, reaches across the ladder, and obtains a new grip on the upper beam (Figure 3.92). (NOTE: On extension ladders, the upper beam of the outermost fly section is grasped.)

Step 4: The firefighters release their hold on the rungs and let the ladder settle to full arm's length, simultaneously pivoting toward the butt (Figure 3.93).

Figure 3.92 Grasp the top beam with the hands closest to the tip end of the ladder.

Figure 3.93 The firefighters release their grip on the rungs with their hands nearest the butt; they pivot and lower the ladder into the final carrying position.

FROM VERTICAL RACKING/REAR REMOVAL

Step 1: The firefighters pull the ladder most of the way out of the rack. The two firefighters position themselves on the bed section side of the ladder (assuming an extension ladder), one at each end (Figure 3.94).

Step 2: The two firefighters grasp the rungs of the ladder with both hands, and lift it from the rack.

Step 3: The two firefighters, without bending over, lower the ladder as far as possible (Figure 3.95).

Step 4: While retaining the grasp on a rung with the hand nearest the butt, each firefighter releases the hold with the other hand, reaches across the ladder, and obtains a new grip on the upper beam (Figure 3.96). (NOTE: On extension ladders, the upper beam of the outermost fly section is grasped.)

Step 5: The firefighters release their hold on the rungs and let the ladder settle to full arm's length, simultaneously pivoting toward the butt.

Figure 3.94 The firefighters position themselves on the bed section side of the ladder at either end.

Figure 3.95 The firefighters lower the ladder without bending.

Figure 3.96 With the hand closest to the tip end, reach over and grasp the top beam.

FROM FLAT RACKING/SIDE REMOVAL

Step 1: The firefighters position themselves facing the ladder, one at each end. Both grasp the beam and pull the ladder outward. As soon as the ladder is part way out, the hands are shifted to grip the rungs (Figure 3.97).

Step 2: The firefighters pull the ladder clear of the rack and lower it. At the same time, the far beam is allowed to swing downward so that the ladder ends up being held vertical.

Step 3: Each firefighter shifts the grip of the hand farthest from the butt to the upper beam. The firefighters then release their grasp on the rung (Figure 3.98).

Step 4: The firefighters simultaneously lower the ladder to arm's length and pivot toward the butt.

Figure 3.97 Pull the ladder out of the rack.

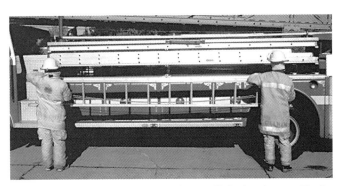
Figure 3.98 Reach over and grasp the beam with the hand nearest the tip.

FROM FLAT RACKING/REAR REMOVAL

Step 1: The two firefighters pull the ladder nearly out of the rack. Both firefighters position themselves on the same side, one at each end.

Step 2: Both firefighters grasp the ladder, slide it clear of the rack, and tilt it to vertical (Figure 3.99).

Step 3: Each firefighter shifts the grasp of the hand farthest from the butt to the upper beam as the hand nearest the butt is released. The firefighters pivot toward the butt and lower the ladder to arm's length.

Figure 3.99 Tilt the ladder toward vertical.

FROM FLAT ON THE GROUND

Step 1: The two firefighters tilt up one beam so that the ladder is resting on the other beam.

Step 2: The two firefighters position themselves on the same side of the ladder (on the bed section side of extension ladders), one near each end. They squat slightly, facing the butt, and grasp the upper beam with the near hand (the beam of the outermost fly section on an extension ladder) (Figure 3.100).

Step 3: The firefighters then stand, lifting the ladder until it is at arm's length.

Figure 3.100 Reach over the ladder and grasp the outside beam.

Special Procedures for Carrying Roof Ladders

Procedures previously described are for ladders that are carried butt forward. In some cases two firefighters will carry a roof ladder with the intention of climbing another ground ladder and placing the roof ladder with hooks deployed on a sloped roof. In this situation, the firefighters should use the low-shoulder method and have the tip (hooks) forward (Figure 3.101). When this situation is the case, the previous instructions regarding firefighters pivoting toward the butt should be reversed so that firefighters pivot toward the tip. The roof ladder, with hooks closed, is carried to the foot of the second ladder. The firefighter at the tip opens the hooks before beginning to climb the second ladder (Figure 3.102).

Figure 3.101 Roof ladders are usually carried with the tip forward.

Figure 3.102 The hooks may be extended before climbing the extension ladder.

Two-Firefighter Multiple-Ladder Carry

When apparatus is unable to gain access to the side of the fire building that requires ladders, firefighters may need to carry more than one ladder at a time. Two firefighters should be able to transport one extension ladder and one single or roof ladder by using either the flat multiple-ladder carry or the arm's length on-edge method.

FLAT MULTIPLE-LADDER CARRY

Step 1: The firefighters remove both ladders from their storage rack and place them flat on the ground.

Step 2: The firefighters place either the single or the roof ladder on top of the extension ladder (Figure 3.103). (NOTE: If the extension ladder was the first ladder removed from the apparatus, the single or the roof ladder may be taken directly from the apparatus and placed onto the extension ladder without having to be placed on the ground.)

Step 3: The firefighters position themselves at each end of the ladders. The firefighter at the butt should have his or her back to the ladders, and the firefighter at the tip should be facing the ladders (Figure 3.104).

Figure 3.103 Place the roof ladder on top of the extension ladder.

Figure 3.104 The firefighter at the butt holds the ladders from behind.

Step 4: The firefighters grab the end rungs of the bottom ladder. The firefighter at the butt should have his or her palms toward the rear. The firefighter at the tip may have his or her palms in either direction.

Step 5: Using their knees to lift, the firefighters lift and carry the ladders in much the same manner as a litter or stretcher would be carried (Figure 3.105).

Figure 3.105 Lift the ladders using your leg muscles.

ARM'S LENGTH ON-EDGE METHOD

Step 1: The firefighters remove the first ladder from the apparatus and lay it flat on the ground, about 4 feet (1.2 m) from the apparatus (Figure 3.106).

Step 2: The firefighters remove the second ladder from the apparatus and set it standing on-edge on the ground (Figure 3.107).

Figure 3.106 Lay the first ladder about 4 feet (1.3 m) from the apparatus.

Figure 3.107 Set the second ladder on the ground on one beam.

Step 3: While balancing the second ladder, the firefighters reach over and tilt the first ladder to a vertical position (Figure 3.108).

Step 4: Grasping the top beam of each ladder, the two firefighters stand up (Figure 3.109).

Figure 3.108 Tilt the first ladder up onto one beam.

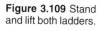

Figure 3.109 Stand and lift both ladders.

Figure 3.110 A completed three-firefighter shoulder carry.

Figure 3.111 The third firefighter stands to the side at one end of the ladder as the other two remove the ladder from the apparatus.

THREE-FIREFIGHTER CARRIES

Three-firefighter carries are typically used on extension ladders of up to 35 feet (11 m). There are four common three-firefighter carries: the flat-shoulder method, the flat arm's length method, the low-shoulder method, and the arm's length on-edge method.

Flat-Shoulder Method

The flat-shoulder carry has two firefighters, one at each end, on one side of the ladder and one firefighter on the other side in the middle (Figure 3.110).

FROM VERTICAL RACKING/SIDE REMOVAL

Step 1: Two firefighters face the ladder, one at each end. They grasp convenient rungs with both hands. The third firefighter, who is not involved in removal of the ladder, takes a position beside the apparatus, adjacent to one end of the ladder (Figure 3.111).

Step 2: The two firefighters grasp the ladder, lift it off the rack, and step back clear of the apparatus (Figure 3.112). (**CAUTION:** Care must be taken before stepping back to be sure that the way is clear and that there are no potholes or other hazards such as previously removed ladders.)

Figure 3.112 The two firefighters lift the ladder free of the apparatus.

Step 3: The third firefighter steps into the space between the apparatus and the ladder and faces the ladder at midpoint (Figure 3.113).

Figure 3.113 The third firefighter assumes a position in the middle of the ladder on the side opposite the other two firefighters.

Step 4: The third firefighter, with assistance from the other two firefighters, lifts the lower beam, pivots toward the butt, and places the beam on his or her shoulder. An alternate method is for the two firefighters to tilt the top beam down toward the third firefighter (Figures 3.114 a through c).

Figure 3.114a The middle firefighter begins to lift the lower beam.

Figure 3.114b The middle firefighter begins to pivot toward the butt.

Figure 3.114c Once the pivot is complete, the ladder is on the middle firefighter's shoulder.

Step 5: With the ladder resting on the third firefighter's shoulder, the other two firefighters pivot toward the butt and lift the near beam onto their shoulders (Figure 3.115).

Figure 3.115 The other two firefighters now pivot toward the butt and shoulder the ladder.

FROM VERTICAL RACKING/REAR REMOVAL

Step 1: One firefighter stands facing the butt and grasps the bottom rung. The other two firefighters stand adjacent to the rear of the apparatus, facing each other on opposite sides of the ladder (Figure 3.116).

Figure 3.116 Two firefighters stand on either side of the ladder as the third pulls the ladder from the rack.

Step 2: The first firefighter, with assistance from the other two firefighters, pulls the ladder out of the rack (Figure 3.117).

Step 3: When the ladder is almost clear of the rack, the first firefighter shifts to one side of the butt, faces the ladder, and grasps two rungs. The firefighter at the tip end, on the same side as the first firefighter, grasps two rungs. The third firefighter, on the side opposite the other two, shifts to the ladder's midpoint and faces the ladder (Figure 3.118).

Step 4: The third firefighter, with assistance from the other two firefighters, lifts the lower beam, pivots toward the butt, and places the beam on his or her shoulder (Figure 3.119).

Step 5: With the ladder resting on the third firefighter's shoulder, the other two firefighters simultaneously pivot toward the butt and place the near beam on their shoulders (Figure 3.120).

Figure 3.120 The firefighters on either end make the pivot and shoulder the ladder.

FROM FLAT RACKING/SIDE REMOVAL

Step 1: Two firefighters stand facing the ladder, one at each end, and grasp the near beam. The third firefighter stands at the side of the apparatus near one end of the ladder (Figure 3.121).

Step 2: The two firefighters slide the ladder part way out of the rack, and each shifts the grip of the hand nearest the tip to a rung. The ladder is then pulled clear of the rack, and the near beam is brought onto their shoulders. The ladder is kept in the horizontal plane (Figure 3.122).

Figure 3.117 The firefighters will begin to spread out as the ladder nears the end of the rack.

Figure 3.118 Two firefighters should be at opposite ends of the ladder on the same side. The third firefighter is in the middle of the ladder on the side opposite the other two firefighters.

Figure 3.121 Two firefighters begin to pull the ladder from the rack as the third waits to the side at one end.

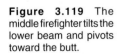

Figure 3.119 The middle firefighter tilts the lower beam and pivots toward the butt.

Figure 3.122 The two firefighters hold the ladder horizontal as the third firefighter moves into position.

Step 3: The third firefighter proceeds into the space between the apparatus and the ladder, moves to the ladder's midpoint, pivots to face the butt, and places the near beam on his or her shoulder (Figure 3.123).

Figure 3.123 The middle firefighter shoulders the ladder.

FROM FLAT RACKING/REAR REMOVAL

Step 1: One firefighter faces the butt. The other two firefighters face each other, one on each side at the butt (Figure 3.124). All three assist in pulling the ladder from the rack (Figure 3.125).

Figure 3.124 One firefighter pulls the ladder from the rack as the other two wait on opposite sides of the ladder.

Figure 3.125 Pull the ladder from the rack.

Step 2: When the ladder is pulled nearly from the rack, the firefighter at the butt shifts to one side. This procedure results in two firefighters being on the same side, one at each end. The third firefighter is positioned at midpoint on the opposite side. All three face the ladder, place their hands palms up under the beams, and lift the ladder clear of the rack (Figure 3.126).

Step 3: The firefighters raise the ladder to shoulder height. Each firefighter then pivots toward the butt and places the beam on his or her shoulder (Figure 3.127).

Figure 3.126 Note that on heavier ladders it may be necessary for the pulling firefighter to face the end of the ladder as opposed to the beam.

Figure 3.127 All firefighters should now shoulder the ladder.

FROM FLAT ON THE GROUND

Step 1:   Two firefighters kneel on one side of the ladder, one at either end, facing the tip. The third firefighter kneels on the opposite side at midpoint, also facing the tip end. In each case, the knee closer to the ladder is the one touching the ground (Figure 3.128).

Step 2:   In the same motion, the firefighters begin to stand and lift the ladder. When the ladder is about chest high, the firefighters pivot toward the butt and place the beam onto their shoulders (Figure 3.129).

Figure 3.130 A completed three-firefighter flat arm's length carry.

Figure 3.128 The firefighters face the tip of the ladder with their knees nearest the ladder on the ground.

Figure 3.129 Raise, pivot toward the butt, and shoulder the ladder.

Step 2:   The two firefighters grasping the ladder lift it off the rack and step back clear of the apparatus. (**CAUTION:** Care must be taken before stepping back to be sure that the way is clear and that there are no potholes or other hazards such as previously removed ladders.)

Step 3:   The third firefighter steps into the space between the apparatus and the ladder and faces the ladder at midpoint.

Step 4:   The third firefighter grasps the lower beam and brings it up so that the ladder approaches horizontal (Figures 3.131 a and b).

Flat Arm's Length Method

The three-firefighter arm's length carry is shown in Figure 3.130. Note that the positioning of the three firefighters is basically the same as the flat-shoulder carry.

FROM VERTICAL RACKING/SIDE REMOVAL

Step 1:   Two firefighters face the ladder, one at each end, and grasp convenient rungs with both hands. The third firefighter, who is not involved in removal of the ladder, takes a position beside the apparatus, adjacent to one end of the ladder.

Figure 3.131a The third firefighter takes a position between the apparatus and the middle of the ladder.

Figure 3.131b Tilt the ladder toward the middle firefighter.

Step 5: Each firefighter shifts the grip of the hand nearest the tip to a convenient rung (Figure 3.132).

Step 6: The firefighters lower the ladder the rest of the way to horizontal at arm's length.

Figure 3.132 The firefighters lower the ladder keeping their hands on the rungs, palms facing the rear.

FROM FLAT RACKING/SIDE REMOVAL

Step 1: Two firefighters stand facing the ladder, one at each end, and grasp the near beam. The third firefighter stands at the side of the apparatus, near one end of the ladder.

Step 2: The two firefighters slide the ladder part way out of the rack. Each shifts the hands to grip two rungs.

Step 3: The firefighters remove the ladder from the rack and tilt it to vertical. Then the two firefighters step back (Figure 3.133). (**CAUTION:** Care must be taken before stepping back to be sure that the way is clear and that there are no potholes or other hazards such as previously removed ladders.)

Figure 3.133 The third firefighter waits at one end as the ladder is pulled from the rack.

Step 4: The third firefighter grasps the lower beam and brings it up so that the ladder approaches horizontal (Figure 3.134).

Step 5: Each firefighter shifts the grip of the hand nearest the tip to a convenient rung (Figure 3.135).

Step 6: The firefighters lower the ladder the rest of the way to horizontal at arm's length.

Figure 3.134 The middle firefighter brings the ladder to the horizontal position.

Figure 3.135 The firefighters adjust their grip from the beam to the rung.

FROM FLAT RACKING/REAR REMOVAL

Steps 1
and 2:   These steps are the same as Steps 1 and
2 for the three-firefighter flat-shoulder
carry from flat racking/rear removal except that the three firefighters grasp the
rungs and lift the ladder clear of the rack.
They then let it settle to arm's length.

FROM FLAT ON THE GROUND

Step 1:   Two firefighters kneel beside the ladder
facing the butt, one at each end. The third
firefighter kneels on the opposite side at
midpoint, also facing the butt. All
firefighters grasp a rung with their near
hands (Figure 3.136).

Step 2:   All firefighters stand and simultaneously
lift the ladder to arm's length.

Figure 3.136 The firefighters kneel facing the butt.

Low-Shoulder Method

The procedures for three firefighters to use the
low-shoulder carry are the same as those used for
the two-firefighter low-shoulder carry except that
the third firefighter is positioned at midpoint (Figure 3.137). All three firefighters should be on the
same side of the ladder. (**NOTE:** In order for three
firefighters to utilize this carry, they should be
nearly the same height.)

Figure 3.137 A completed three-firefighter low-shoulder carry.

Arm's Length On-Edge Method

The procedures for three firefighters to use the
arm's length on-edge carry are the same as for the
two-firefighter arm's length on-edge carry except
that the third firefighter is positioned at midpoint
between the other two firefighters, all on the same
side of the ladder (Figure 3.138).

Figure 3.138 A completed three-firefighter arm's length on-
edge carry.

FOUR-FIREFIGHTER CARRIES

The same four methods used by three
firefighters for carrying ladders are used by four
firefighters except that there is a change in the
positioning of the firefighters to accommodate the
fourth firefighter.

Flat-Shoulder Method

When four firefighters use the flat-shoulder
carry, two are positioned at each end of the ladder,
opposite each other (Figure 3.139).

Figure 3.139 A completed four-firefighter flat-shoulder carry.

PROCEDURE FOR FLAT OVERHEAD RACKING ON A PUMPER/REAR REMOVAL

The steps for the four-firefighter flat-shoulder carry from a pumper overhead rack are as follows:

Step 1: Two firefighters stand on the rear step or on intermediate steps, reach up, and grasp a convenient rung with their near hands while maintaining a grasp on the apparatus with their other hands.

Step 2: The third and fourth firefighters stand facing the rear of the apparatus, approximately 10 feet (3 m) from the tailboard (Figure 3.140).

Figure 3.140 Two firefighters prepare to remove the ladder while the other two wait on the ground.

Step 3: Firefighters 1 and 2 slide the ladder back (Figure 3.141).

Step 4: The waiting firefighters grasp the butt of the ladder when the ladder passes its balance point and the butt tilts downward (Figure 3.142).

Figure 3.141 The firefighters on the ground watch as the ladder begins to be moved from the bed.

Figure 3.142 The firefighters on the ground catch the end of the ladder as it is lowered to them.

Step 5: Firefighters 1 and 2 continue sliding the ladder back until just the tip is resting on the lip of the rack (Figure 3.143). Firefighters 3 and 4 step back as the ladder is being slid back (Figure 3.144).

Figure 3.143 With the tip of the ladder resting on the end of the rack, the two firefighters on the truck reposition to continue the operation.

Figure 3.144 The firefighters on the ground should catch the ladder in a comfortable position. If possible, it should be shouldered immediately.

Figure 3.146 All firefighters shoulder the ladder, facing the butt.

Step 6: Firefighters 1 and 2 release their grip, leaving the tip of the ladder resting on the lip of the rack. They then descend to the tailboard, turn and face the butt, and reach up and grasp the ladder beam (Figure 3.145).

Flat Arm's Length Method

When the flat arm's length carry is used by four firefighters, the firefighters are positioned exactly the same as they are for the flat-shoulder carry except that the ladder is carried at arm's length (Figure 3.147).

Figure 3.145 The two firefighters on the truck move to the tailboard and grasp the ladder.

Figure 3.147 A completed four-firefighter flat arm's length carry.

Low-Shoulder Method

This carry is not recommended for use with pole ladders. When used to carry an extension ladder, the four firefighters are spaced evenly along the length of the ladder (Figure 3.148). The evolution is performed in the same manner as it is for the two-firefighter low-shoulder carry.

Step 7: Firefighters 1 and 2 lift the ladder from the rack, bring it down onto their shoulders, and step from the tailboard to the ground. Firefighters 3 and 4 pivot 180 degrees to complete positioning for the carry (Figure 3.146). (**CAUTION:** If the tailboard is too high off the ground to safely step down while shouldering the ladder, firefighters 1 and 2 should set their end of the ladder down on the tailboard, step down to the ground, and then shoulder the ladder again.)

Arm's Length On-Edge Method

This carry is also not recommended for use with pole ladders. When used to carry an extension ladder, the four firefighters are spaced evenly along the length of the ladder, the same as for the low-shoulder method (Figure 3.149). The evolution is performed in the same manner as it is for the two-firefighter arm's length on-edge method. All the firefighters should be on the same side (base section side) of the ladder.

Figure 3.148 A completed four-firefighter low-shoulder carry.

Figure 3.149 A completed four-firefighter arm's length on-edge carry.

FIVE-FIREFIGHTER CARRIES

Five firefighters are normally required for a 40-foot (12 m) or longer pole ladder (Figure 3.150). Manning levels are often such that fewer than five firefighters respond with the apparatus. Because of this situation, some fire departments do not make an effort to use pole ladders and have stopped drilling with them. This was one of the primary

Figure 3.150 When possible, at least five firefighters should be used to carry a pole ladder.

reasons that led the NFPA 1904 committee to drop these ladders from the list of those required to be carried on aerial apparatus. However, consideration must be given to the fact that when aerial apparatus cannot gain access to a particular area, the pole ladder may be the only piece of equipment available to get the job done. If this situation happens, personnel will have to be recruited from other crews, making it important that all fire fighting personnel be familiar with handling pole ladders.

Due to the weight, bulk, and the presence of staypoles, the flat-shoulder and flat arm's length carries are used.

Flat-Shoulder Method

The same procedures are used as those for the three-firefighter flat-shoulder method except that the firefighters are positioned differently. Three are on one side, evenly spaced down the length of the ladder, and two are spaced evenly down the other side halfway between the three firefighters on the opposite side (Figure 3.151).

Figure 3.151 A completed five-firefighter flat-shoulder carry of a pole ladder.

PROCEDURE FOR FLAT OVERHEAD RACKING ON A PUMPER/REAR REMOVAL

The same basic procedure is used as that for the four-firefighter flat overhead racking on a pumper/rear removal except that the firefighters on the ground are spaced differently. Initially, all three firefighters on the ground stand where the butt will come down — one at the end and one on each side

(Figure 3.152). When the ladder is almost out of the rack, the firefighters on the sides begin to shift their positions (Figure 3.153). They should be in place by the time the two firefighters on the back of the apparatus are ready to step down (Figure 3.154). The firefighter at the butt shifts to the outside as the evolution is completed (Figure 3.155).

Figure 3.152 Three firefighters wait for the ladder to be lowered to them on the ground.

Figure 3.153 The ground firefighters spread out as the ladder is lowered to them.

Figure 3.154 The firefighters on the apparatus move to the tailboard so that the evolution may be continued.

Figure 3.155 The firefighters that were on the apparatus step down, and the ladder may be carried to the needed location.

Flat Arm's Length Method

The same procedures are used as those for the three-firefighter flat arm's length carry except that three firefighters are positioned on one side — one at the tip, one at midpoint, and one at the butt. The other two firefighters are positioned on the opposite side — one halfway between the tip and midpoint, and one halfway between the midpoint and the butt (Figure 3.156).

Figure 3.156 A completed five-firefighter flat arm's length carry.

SIX-FIREFIGHTER CARRIES

The ideal crew size for a pole ladder is six firefighters. With this number of individuals lifting and carrying, the risk of straining is greatly reduced.

The procedures for six firefighters using the flat-shoulder carry are the same as those described in detail in the section on the three-firefighter flat-shoulder carry except for the positioning of the firefighters. They are located at the butt, midpoint, and tip on each side (Figure 3.157). Procedures for the six-firefighter flat arm's length carry are the same, except that the ladder is lowered to arm's length rather than being placed on the shoulder (Figure 3.158).

Figure 3.157 A completed six-firefighter flat-shoulder carry.

Figure 3.158 A completed six-firefighter flat arm's length carry.

Figure 3.160 Use both hands to carry a folding ladder.

The procedure for removing a ladder from flat overhead racking on a pumper with six firefighters is the same as that for four firefighters except that there are four firefighters, instead of two, on the ground to receive the ladder as it is tilted down from the rack (Figure 3.159). Two firefighters assume a position at midpoint for the carry.

Figure 3.159 The initial positioning for unloading a pole ladder from an overhead rack with six firefighters.

CARRYING OTHER LADDERS

The procedures for carrying some of the specialized ladders listed in Chapter 1 are explained in the following sections.

Folding Ladders

Folding ladders should always be carried in the closed position with the tip forward. Hold the ladder horizontal while grasping with both hands (Figure 3.160).

Combination Ladders

Combination ladders are usually carried by one firefighter. The arm's length on-edge method is used (Figure 3.161).

Pompier Ladders

Pompier ladders are carried by one firefighter using a shoulder carry (Figure 3.162).

Figure 3.161 Combination ladders are usually carried at arm's length on-edge.

Figure 3.162 The proper method for carrying a pompier ladder.

Special Carry For Narrow Passageways

When firefighters using either the flat-shoulder carry or the flat-arm's length carry encounter a narrow passageway, they will need to adjust their carry to make it through. An easy way to overcome this problem is for the firefighters to raise the ladder over their heads, assuming that there is space overhead. The procedure is as follows:

Step 1: The firefighter at the right tip position gives the signal to shift to the overhead carry.

Step 2: All firefighters lift the ladder until their arms are extended straight overhead.

Step 3: The firefighter at the left butt steps back and under the ladder. The firefighter at the left tip steps forward and under the ladder.

Step 4: The firefighters on the right side shift under the ladder (Figures 3.163 a and b).

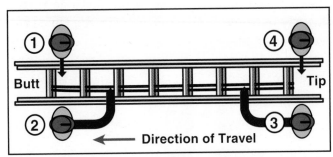

Figure 3.163a This diagram indicates the shifting movements of the firefighters in order to negotiate a narrow passage.

Figure 3.163b A completed narrow passage carry.

POSITIONING (PLACEMENT) OF GROUND LADDERS

Proper positioning, or placement, of ground ladders is important because it affects the safety and efficiency of operations. This section contains some of the basic considerations and requirements for ground ladder placement.

Responsibility For Positioning

Normally, an officer will designate the general location where the ladder is to be positioned and/or the task is to be performed. However, personnel carrying the ladder frequently decide on the exact spot where the butt is to be placed. The firefighter nearest the butt is the logical person to make this decision because this end is placed on the ground to initiate raising the ladder. Where there are two firefighters at the butt, as in the four- and six-firefighter flat carries, the one on the right side is usually the one responsible for placement (Figure 3.164). However, this designation is an option as far as each department's policy is concerned.

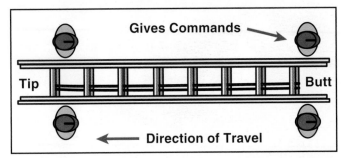

Figure 3.164 The firefighter on the right side of the butt gives the commands.

Factors Affecting Ground Ladder Placement

When placing ladders, there are two objectives to be met: first, to place the ladder properly for its intended use, and second, to place the butt the proper distance from the building for safe and easy climbing. There are numerous factors that dictate the exact place to position the ladder.

If the ladder is to be used to provide a vantage point from which a firefighter can break out a window for ventilation, it should be placed alongside the window to the windward (upwind) side. The tip should be about even with the top of the window (Figure 3.165). The same position can be used when firefighters desire to climb into or out of narrow windows.

If the ladder is to be used for entry or rescue from a window, usually the ladder tip is placed even with or slightly below the sill (Figure 3.166). If the sill projects out from the wall, the tip of the ladder can be wedged up under the sill for additional stability (Figure 3.167). If the window opening is wide enough to permit the ladder tip to project into it and still allow room beside it to facilitate entry and rescue, the ladder should be placed so that two or three rungs extend above the sill (Figure 3.168).

Figure 3.165 Place the tip adjacent to the top of the window opening.

Figure 3.166 For rescue, place the tip just below the lower sill.

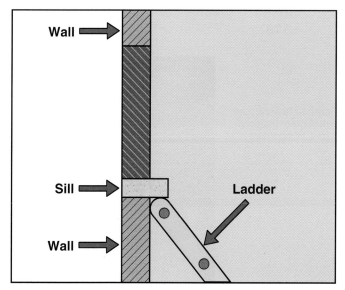

Figure 3.167 Wedging the tip under the sill makes for a more stable placement.

Figure 3.168 In wide windows, the ladder may be extended into one side of the opening.

When a ladder is to be used as a vantage point from which to direct a hose stream into a window opening and no entry is to be made, it is raised directly in front of the window, with the tip on the wall above the window opening (Figure 3.169). Care must be taken to keep flames from engulfing the tip of the ladder. If this situation cannot be avoided, the ladder is raised just to the sill.

Figure 3.169 Place the ladder directly over the opening so that a hose stream may be discharged into the window.

Other placement guidelines include the following:

- Roofs should be laddered from at least two points on opposite corners of the building (Figure 3.170).

Figure 3.170 The roof, or any portions of the building, should be laddered from at least two points. Here the upper story is laddered from two points. *Courtesy of Bill Tompkins.*

- Care should be taken to avoid placing ladders over horizontal openings such as windows and doors.

- Ladder placement should take advantage of strong points in building construction.

- When a ladder is to be used as a support for a smoke ejector, it is raised directly in front of the window, with the tip on the wall above the window opening.

- Avoid placing ladders where they may come into contact with overhead obstructions, such as wires, tree limbs, or signs, when they are raised (Figure 3.171).

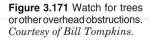

Figure 3.171 Watch for trees or other overhead obstructions. *Courtesy of Bill Tompkins.*

- Avoid placing ladders on uneven terrain or on soft spots such as mud holes.

- Avoid placing ladders on main paths of travel that firefighters or evacuees will need to use (Figure 3.172).

Figure 3.172 Do not place ladders in front of doors.

- Avoid placing ladders where they may contact either burning surfaces or openings with flames present.

- Avoid placing ladders on top of sidewalk elevator trapdoors or sidewalk deadlights. These areas may give way under the weight of both the ladder and firefighters (Figure 3.173).

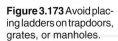

Figure 3.173 Avoid placing ladders on trapdoors, grates, or manholes.

• Do not place ladders where they will have to be raised against unstable walls or surfaces.

When the ladder has been raised and lowered into place, the desired angle of inclination is approximately 75 degrees (Figure 3.174). This angle provides good stability and places stresses on the ladder properly. It also provides for easy climbing, because it permits the climber to stand perpendicular at arm's length from the rungs.

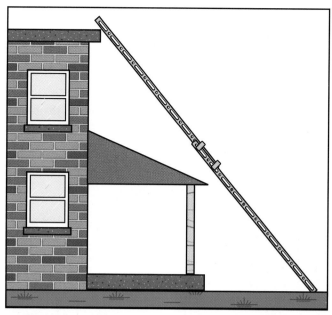

Figure **3.175** Sometimes obstructions force ladders to be placed at less than optimal positions.

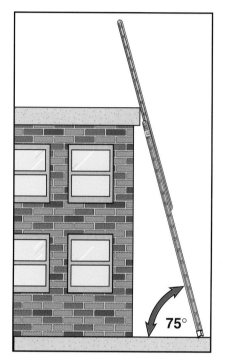

Figure 3.174 Ladders should be placed at a 75-degree angle.

If the butt of the ladder is placed too far away from the building, the load carrying capacity of the ladder is reduced and it has more of a tendency to slip. Placement at such an angle may be necessary, however, when there is an overhang on the building (Figure 3.175). Either tie the bottom of the ladder off or heel it at all times when it is placed at a poor angle.

If the butt is placed too close to the building, its stability is reduced because climbing tends to cause the tip to pull away from the building.

An easy way to determine the proper distance between the heel of the ladder and the building is to divide the used length of ladder by four. For example, if 20 feet (6 m) of ladder is needed to reach a window, the butt should be placed 5 feet (1.5 m) out from the building (20 feet divided by 4 [6 m divided by 1.5]) (Figure 3.176). Exact measurements are unnecessary on the fire scene. Firefighters develop

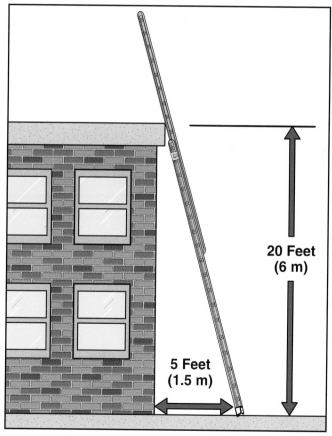

Figure 3.176 A ladder that is raised 20 feet (4.8 m) should have the base 5 feet (1.6 m) from the building.

the experience to judge visually the proper positioning for the ladder. The proper angle can also be checked by standing on the bottom rung and reaching out for the rung in front of the firefighter. The firefighter should be able to grab the rung, while standing straight up, with arms extended straight out in front (Figure 3.177). Newer ladders are equipped with an inclination marking on the outside of the beam whose lines become perfectly vertical and horizontal when the ladder is properly set (Figure 3.178).

Figure 3.177 Check for a proper angle by standing on the bottom rung and reaching for the rung directly at shoulder level.

Figure 3.178 Newer ladders have labels that help firefighters achieve a proper climbing angle.

Chapter 3 Review

Vocabulary

Be sure that you know the chapter-related meanings of the following words.

- handling ladders *(57)*
- ladder reach *(62)*
- designated ladder length *(62)*
- reach factor *(63)*
- angle of inclination *(99)*

Application Of Knowledge

1. Compute the following problems:
   - You are directed to ladder the eaves of a two-story residence. Assuming that each story is 8 feet (2.5 m), what ladder length will be required? *(63)*
   - You are directed to place a ladder for access into a third floor window at a boarding house. Assuming that each story is 10 feet (3 m), what ladder length will be required? *(63)*
   - What length of ladder will be required to perform a window rescue from a fourth floor window in a commercial occupancy? *(63)*

2. Examine your department's ladder release mechanisms to determine how they operate. *(local protocol)*

3. Examine your department's ladder carrying apparatus and then answer the following questions: *(local protocol)*
   - What ladders (types and lengths) are carried and where are they carried on the apparatus?
   - Are the ladders racked with the butt toward the front or toward the rear of the apparatus?
   - Where ladders are nested together, can one ladder be removed leaving the other(s) secured in place? (In particular can the roof ladder be removed from the side of the pumper and leave the extension ladder secured?)
   - In what order do the ladders that nest together rack? (Pumper extension ladder goes on first, roof ladder second, or vice versa?)
   - Is the top fly of the extension ladder on the inside or on the outside when the ladder is racked on the side of the apparatus?
   - How are the ladders secured?
   - When ladders are mounted vertically on the side of apparatus, which rungs go in or near the brackets? Does your department mark its ladders to indicate when rungs go in or near the brackets?

4. Choose from the following ladder carries those that are appropriate to your department and equipment, or ask your training officer to choose appropriate carries. Practice the chosen carries under your training officer's supervision.
   - Perform the one-firefighter low-shoulder carry from vertical racking/side removal. *(66)*
   - Perform the one-firefighter low-shoulder carry from vertical racking/rear removal. *(66)*
   - Perform the one-firefighter low-shoulder carry from flat racking/side removal. *(68)*
   - Perform the one-firefighter low-shoulder carry from flat racking/rear removal. *(68)*
   - Perform the one-firefighter low-shoulder carry from flat on the ground. *(68)*
   - Perform the one-firefighter high-shoulder carry from vertical racking/side removal. *(69)*
   - Perform the one-firefighter high-shoulder carry from vertical racking/rear removal. *(69)*
   - Perform the one-firefighter high-shoulder carry from flat racking/side removal. *(70)*
   - Perform the one-firefighter high-shoulder carry from flat racking/rear removal. *(70)*
   - Perform the one-firefighter high-shoulder carry from flat on the ground. *(70)*
   - Perform the one-firefighter arm's length carry from vertical racking/side removal. *(71)*

- Perform the one-firefighter arm's length carry from vertical racking/rear removal. *(71)*
- Perform the one-firefighter arm's length carry from flat racking/side removal. *(72)*
- Perform the one-firefighter arm's length carry from flat racking/rear removal. *(72)*
- Perform the one-firefighter arm's length carry from flat on the ground. *(73)*
- Perform the one-firefighter roof ladder carry. *(73, 74)*
- Perform the two-, three-, and four-firefighter low-shoulder carry from vertical racking/side removal. *(75)*
- Perform the two-, three-, and four-firefighter low-shoulder carry from vertical racking/rear removal. *(75)*
- Perform the two-, three-, and four-firefighter low-shoulder carry from flat racking/side removal. *(76)*
- Perform the two-, three-, and four-firefighter low-shoulder carry from flat racking/rear removal. *(76)*
- Perform the two-, three-, and four-firefighter low-shoulder carry from flat on the ground. *(76)*
- Perform the two-firefighter hip or underarm carry from vertical racking/side removal. *(77)*
- Perform the two-firefighter hip or underarm carry from vertical racking/rear removal. *(77)*
- Perform the two-firefighter hip or underarm carry from flat racking/side removal. *(78)*
- Perform the two-firefighter hip or underarm carry from flat racking/rear removal. *(79)*
- Perform the two-firefighter hip or underarm carry from flat on the ground. *(79)*
- Perform the two-, three-, and four-firefighter arm's length on-edge carry from vertical racking/side removal. *(80, 90, 92)*
- Perform the two- and four-firefighter arm's length on-edge carry from vertical racking/rear removal. *(80, 92)*
- Perform the two-, three-, and four-firefighter arm's length on-edge carry from flat racking/side removal. *(81, 90, 92)*
- Perform the two-, three-, and four-firefighter arm's length on-edge carry from flat racking/rear removal. *(81, 90, 92)*
- Perform the two-, three-, and four-firefighter arm's length on-edge carry from flat on the ground. *(82, 90, 92)*
- Perform the two-firefighter roof ladder carry. *(82)*
- Perform the two-firefighter flat multiple-ladder carry. *(83)*
- Perform the two-firefighter flat multiple-ladder arm's length on-edge carry. *(83)*
- Perform the three- and six-firefighter flat-shoulder carry from vertical racking/side removal. *(84, 94)*
- Perform the three- and six-firefighter flat-shoulder carry from vertical racking/rear removal. *(85, 94)*
- Perform the three- and six-firefighter flat-shoulder carry from flat racking/side removal. *(86, 94)*
- Perform the three- and six-firefighter flat-shoulder carry from flat racking/rear removal. *(87, 94)*
- Perform the three- and six-firefighter flat-shoulder carry from flat on the ground. *(88, 94)*
- Perform the four- and five-firefighter flat-shoulder or flat arm's length carry from flat overhead pumper racking/rear removal. *(90, 93)*
- Carry a folding ladder. *(95)*
- Carry a combination ladder. *(95)*
- Perform the narrow passageway carry. *(96)*

Review Activities

1. List factors that determine the location and method of mounting ladders on apparatus. *(57)*
2. Describe two rules of thumb for locating ladders on apparatus. *(57)*
3. Name the type and length of ladders required by NFPA 1901 to be carried on all pumper fire apparatus. *(57)*
4. Answer the following questions:
   - How are ladders usually mounted on pumpers? *(58)*
   - What two factors generally dictate that pumper ladders be mounted in an overhead rack? *(58)*
5. Name the type and length of ladders required by NFPA 1904 to be carried on aerial apparatus. *(59)*
6. Describe three racking arrangements commonly used for ground ladders on aerial apparatus. *(59)*
7. Describe the way in which folding ladders and combination ladders are commonly carried on aerial apparatus. *(60)*
8. Describe the NFPA 1902 ladder requirements for each of the following initial attack apparatus:
   - GVW of less than 15,000 pounds (6 810 kg) *(61)*
   - GVW between 15,000 and 20,000 pounds (6 810 kg and 9 080 kg) *(61)*
   - GVW in excess of 20,000 pounds (9 080 kg) *(61)*
9. Describe the NFPA 1903 ladder requirements for mobile water supply apparatus. *(61)*
10. Explain the present ladder requirements (or lack of them) for rescue or squad apparatus. *(61)*
11. List rough measurements for the following distances: *(61)*
    - one residential story
    - residential floor to windowsill
    - one commercial story
    - commercial floor to windowsill
12. Explain the working rules for ladder length/position for the following situations: *(62)*
    - general extension beyond roof edge
    - access to side of window or for ventilation
    - rescue from a window opening
13. Determine and list the ladder length needed for each of the following elevations: *(61)*
    - first story roof
    - second story window
    - second story roof
    - third story window or roof
    - fourth story roof

14. Explain the differences between a ladder's measured length and its designated length, and explain why neither indicates the ladder's reach. *(62)*

15. State the rules of thumb for ladders over and under 35 feet (11 m) in regard to designated length. *(62)*

16. Explain why a 35-foot (11 m) extension ladder is the most versatile of all extension ladders. *(63)*

17. Explain why ladders carried on aerial apparatus are sometimes preferred by firefighters. *(63, 64)*

18. List general considerations for racking, removing, and returning ladders to apparatus. *(64)*

19. Describe safety guidelines for removing ladders from aerial apparatus while stabilizers are being deployed. *(65)*

20. State the warning about removing ladders from aerial apparatus in contact with live electrical equipment. *(65)*

21. List the recommended procedures for safe lifting and lowering techniques. *(65, 66)*

22. Explain special procedures for carrying roof ladders. *(73, 74)*
    * butt position
    * pivot direction
    * hooks (open/closed)
    * tip position

23. List the two general objectives that must be met when placing ground ladders. *(96)*

24. State where the ladder should be positioned in the following instances:
    * Providing a vantage point from which a firefighter can break out a window for ventilation *(96)*
    * Entry or rescue from a window *(96, 97)*
    * Providing a vantage point from which to direct a hose stream *(97)*

25. List general guidelines for placing ladders. *(98)*

26. List the advantages of placing the ladder at a 75-degree angle of inclination. *(99)*

27. List the hazards and disadvantages of placing the ladder butt too close or too far away from the building. *(99)*

28. Write the formula for calculating the proper distance between the heel of the ladder and the building. *(99)*

--- Questions And Notes ---

Chapter 4

Raising Ground Ladders

LEARNING OBJECTIVES

This chapter provides information that will assist the reader in meeting the objectives contained in the Ladders section of NFPA 1001, *Standard for Fire Fighter Professional Qualifications* (1992 edition). The objectives contained in this chapter are as follows:

Fire Fighter I
3-11.2   Carry, position, raise, and lower the following ground ladders:
  (a) 14-ft (4.3-m) single or wall ladder
  (b) 24-ft (7.3-m) extension ladder
  (c) 35-ft (10.7-m) extension ladder
  (d) Folding/attic ladder

Raising Ground Ladders

In the previous chapter, we covered how to remove ground ladders from the apparatus and transport them to the place of use. In this chapter, we look at the next step in the deployment process: raising the ladder to the point where it is needed. As with carrying the ladder, the process of raising the ladder requires the use of sound techniques so that firefighters do not injure themselves. In situations where two or more firefighters are involved in the raise, closely coordinated teamwork is required to assure a smooth operation.

This chapter contains instructions for raising single, extension, and other types of ladders. Included are procedures for single- or multiple-firefighter raises. Also contained in this chapter is information on safe ways of moving a ladder once it has been extended or raised.

In most cases, more than one option is provided; the one best suited for the particular situation should be used. Procedures for individual fire departments may vary from those presented. This situation is acceptable as long as the variation is safe and teamwork results.

RAISING SINGLE AND EXTENSION LADDERS

The methods and precautions for raising single and extension ladders are much the same. With the exception of pole ladders, it is not necessary to place the ladder flat on the ground prior to raising; only the butt need be placed on the ground (Figure 4.1). The transition from carry to raise can and should be smooth and continuous.

This chapter contains step-by-step information only for raising ladders. In every case, the procedure for lowering the ladder will be to reverse the listed steps in the given order.

Figure 4.1 With the exception of pole ladders, it is not necessary to place the ladder flat on the ground prior to raising.

Things To Consider Before Raising The Ladder

Before raising a ladder, there are a number of things firefighters need to consider and precautions they must take. Some of the more important ones are contained in the following sections.

ELECTRICAL HAZARDS

A major concern when raising ladders is contact with live electrical wires or equipment, either by the ladder or by the person who will have to climb it. This danger with metal ladders has been stressed previously. However, many firefighters do not realize that WET wood or fiberglass ladders present the same hazard. To avoid this hazard, care must be taken BEFORE BEGINNING A RAISE (Figure 4.2).

Firefighters need to look overhead for electrical wires or equipment before making the final selection as to where to place a ladder or what method to

Figure 4.2 Always check for electrical hazards before raising a ladder.

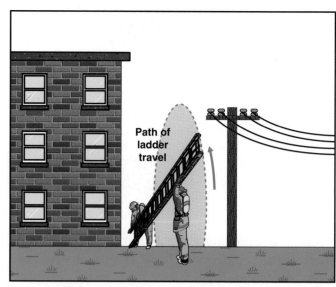

Figure 4.4 By using a parallel raise, the overhead wires are avoided.

use for raising it. IFSTA recommends that all ground and aerial ladders maintain a safe distance of at least 10 feet (3 m) from all energized electrical lines or equipment. This distance must be maintained at all times, including during the raise itself. In some cases, the ladder will come to rest a safe distance from the electrical equipment; however, it will come too close to this equipment during the actual raise (Figure 4.3). In these cases, an alternate method for raising the ladder, such as raising parallel to the building as opposed to perpendicular (discussed later in this chapter), may be required (Figure 4.4).

Figure 4.3 If this ladder is raised perpendicular to the building, it may come into contact with the wires.

FLY POSITION ON EXTENSION LADDERS

The question of whether the fly on an extension ladder should be in (next to the building) or out (away from the building) must also be addressed before starting the discussion of raises. This question has been a matter of controversy in the fire service for many years.

Each ladder manufacturer will specify whether its ladder should be placed with the fly in or out. This recommendation is based on the design of the ladder and the fly position at which tests show it to be strongest. Failure to follow this recommendation could void the warranty of the ladder should a failure or damage occur.

In general, all modern metal and fiberglass ladders are designed to be used with the FLY OUT (away from the building) (Figures 4.5 a and b). Wood ladders that are designed with the rungs mounted in the top truss rail (the only type of wood ladder still manufactured today) are intended to be deployed with the FLY IN (Figure 4.6). Again, consult the manufacturer of your ladders to find out, for certain, the correct fly position.

Some departments have ladders that are intended to be used with the fly out, but prefer the firefighter extending the halyard to be on the outside of the ladder. In this case, firefighters will need to pivot or roll the ladder 180 degrees after it has been extended. Directions for pivoting and rolling ladders safely are given later in this chapter.

Figure 4.5a Metal extension ladders are deployed with the fly out (away from the building).

Figure 4.5b Fiberglass extension ladders are deployed with the fly out (away from the building).

Figure 4.6 Wood ladders are deployed with the fly in (toward the building).

PARALLEL/PERPENDICULAR RAISES

Before starting the raise, firefighters have two choices for the starting position of the ladder: perpendicular (at a right angle) to the building or parallel to the building. The perpendicular position should always be the first preference, because the ladder will ultimately have to be brought to this position to be placed against the building (Figure 4.7).

Figure 4.7 A perpendicular raise.

The parallel raising position is used when circumstances, such as obstructions, prevent the use of the perpendicular raise (Figure 4.8). When this happens, the ladder is brought to vertical and then pivoted to the right-angle position (Figure 4.9). Instructions on how to properly pivot a ladder are given later in this chapter.

Figure 4.8 A parallel raise.

Figure 4.9 It may be necessary to pivot the ladder after a parallel raise.

One-Firefighter Raises

One firefighter may safely raise single ladders and small extension ladders. The following sections show the procedures used to accomplish these raises.

SINGLE LADDER RAISE

Single and roof ladders of 14 feet (4.3 m) or less are easily raised by one firefighter. The firefighter can usually place the butt at the point where it will be located for climbing without heeling against the building or another object before raising. The following steps should be used:

Step 1: Lower the butt to the ground at the proper distance from the building for climbing (Figure 4.10).

Figure 4.10 Place the butt on the ground at an appropriate distance from the building.

Step 2: In one motion, raise the ladder to a vertical position (Figure 4.11).

Step 3: Grasp both beams, heel the butt of the ladder, and lower into the objective (Figure 4.12).

Figure 4.11 Bring the ladder to vertical.   **Figure 4.12** Lower the ladder to the objective.

On single ladders longer than 14 feet (4.3 m), the butt should be placed against the building to heel the ladder as it is raised (Figure 4.13). Once vertical, the firefighter can pull the butt out away from the building to the proper distance for a good climbing angle (Figure 4.14).

Figure 4.13 As the ladder is raised, place the butt against the side of the building.

Figure 4.14 Pull the butt out to achieve a proper climbing angle.

Figure 4.15 Place the butt at a point that will allow for a good climbing angle.

EXTENSION LADDER RAISES

There are two common methods of raising extension ladders with one firefighter: from the high-shoulder carry and from the low-shoulder carry. The type of raise used will depend on the carrying method, the weight of the ladder, and the strength of the firefighter. When raising the ladder from the high-shoulder carry, the following procedure is used:

Step 1: Lower the butt to the ground at a point determined for the proper climbing angle (Figure 4.15). If the ladder will have to be rolled over to get the fly in the proper position, the initial placement should be one ladder width to the side of the final desired location.

CAUTION: Visually check the area overhead for obstructions before bringing the ladder to a vertical position.

Step 2: As the ladder is brought to a vertical position, pivot the ladder 90 degrees, and take a position facing the ladder on the side away from the building (Figure 4.16). (NOTE: If needed, refer to more detailed directions on performing the pivot given later in this chapter.)

Figure 4.16 Bring the ladder to vertical.

Step 3: To extend the ladder, place one foot at the butt of one beam. With the instep, knee, and leg, steady the ladder (Figure 4.17).

Step 4: Grasp the halyard, and extend the fly section with a hand-over-hand motion (Figure 4.18). When the tip is at the desired elevation, make sure that the ladder locks are in place.

Step 5: To lower the ladder into position, place at least one foot either against a butt spur or on the bottom rung while grasping the beams. Lower the ladder gently into the building (Figure 4.19).

Step 6: To position the ladder so that the fly section is facing out, the ladder must be rolled over. Roll the ladder by heeling the outside of the beam on which the ladder will be rolled. Roll the opposite beam over the top of the beam that is being heeled until the ladder is 90 degrees to the building. Shift the handholds and heeling foot, and complete the roll (Figures 4.20 a through c).

Figure 4.20a Prepare to roll the ladder to get the fly out.

Figure 4.17 Brace the ladder using your instep, knee, and leg.

Figure 4.20b Swing the ladder on one beam.

Figure 4.20c Complete the roll.

Figure 4.18 Extend the fly to the necessary height.

Figure 4.19 Lower the ladder into the building.

The major difference in using the one-firefighter raise from the low-shoulder carry is the placement of the butt. In this instance, the building that is being laddered is used to heel the ladder to prevent the ladder butt from slipping while the ladder is brought to the vertical position. When raising the ladder from the low-shoulder carry, the following procedure is used:

Step 1: Place the butt of the ladder on the ground with one of the butt spurs against the wall of the building (Figure 4.21).

Figure 4.21 Place the butt against the building.

Step 2: With your free hand, grasp a rung in front of your shoulder while removing your opposite arm from between the rungs.

Step 3: Step beneath the ladder and grasp a convenient rung with your other hand (Figure 4.22). At this point, the ladder should be flat, with both butt spurs against the building.

Figure 4.22 Swing underneath the ladder.

CAUTION: Visually check the area overhead for obstructions before bringing the ladder to a vertical position. Also visually check the terrain in front of you before stepping forward.

Step 4: Advance hand-over-hand down the rungs toward the butt until the ladder is in a vertical position (Figure 4.23).

Step 5: To extend the ladder, pull the halyard until the ladder has been raised to the desired level and the pawls are engaged (Figure 4.24). Care must be taken to pull straight down on the halyard so that the ladder is not pulled over. During the pull, brace one beam of the ladder with your instep, knee, and leg.

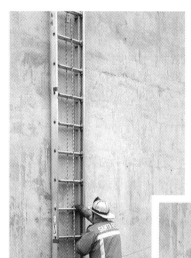

Figure 4.23 Advance down the ladder using the rungs.

Figure 4.24 Extend the fly while bracing the ladder with your leg.

Step 6: To position the ladder for climbing, push against an upper rung to keep the ladder against the building. Grasp a lower rung with your other hand and carefully move the ladder butt out from the building to the desired location (Figure 4.25). If necessary, roll the ladder to bring the fly to the out position using the same technique as described in Step 6 of raising from a high-shoulder carry.

Figure 4.25 Pull the butt out to make a proper climbing angle.

Two-Firefighter Raises

Space permitting, it makes little difference if a ladder is raised parallel with or perpendicular to a building. If raised parallel with the building, the ladder can always be pivoted after it is in the vertical position. Whenever two or more firefighters are involved in raising a ladder, the firefighter at the butt, called the heeler, is responsible for placing it at the desired distance from the building and determining whether the ladder will be raised parallel with or perpendicular to the building. There are two basic ways for two firefighters to raise a ladder: the flat raise and the beam raise. The following are step-by-step procedures for completing each raise:

Step 1: When the desired location for the raise has been reached, the heeler places the ladder butt on the ground while the firefighter at the tip rests the ladder beam on a shoulder (Figure 4.26).

Figure 4.26 The firefighter at the butt places it on the ground.

Step 2: The heeler heels the ladder by either standing on the bottom rung or by placing the toes or insteps on the beams. The heeler then crouches down to grasp a convenient rung or the beams with both hands, and leans back. The firefighter at the tip steps beneath the ladder and grasps a convenient rung with both hands (Figure 4.27).

CAUTION: Visually check the area overhead for obstructions before bringing the ladder to a vertical position. Before stepping forward, visually check the terrain.

Figure 4.27 The firefighter at the butt heels the ladder while the firefighter at the tip swings underneath it.

Step 3: The firefighter at the tip advances hand-over-hand down either the rungs or the beams toward the butt until the ladder is in a vertical position (Figure 4.28). As the ladder comes to a vertical position, the heeler either grasps successively higher rungs or grasps higher on the beams until he or she is standing upright.

Figure 4.28 The tip firefighter advances down the ladder on the rungs to bring the ladder to vertical.

Step 4: Both firefighters face each other and heel the ladder by placing their toes against the same beam. When raising extension ladders, pivot the ladder to position the fly away from the building (fly in for wood ladders) if it is not already in that position. The firefighter on the halyard side of the ladder grasps the halyard and extends the fly section with a hand-over-hand motion (Figure 4.29). When the tip is at the desired elevation, the firefighters make sure the ladder locks are in place.

Step 5: To lower the ladder into position, the firefighter on the outside of the ladder places one foot either against a butt spur or on the bottom rung and grasps the

beams. Both firefighters gently lower the ladder into the building (Figure 4.30). If the ladder has not yet been turned to position the fly in the out position, it can be done at this time.

Figure 4.29 The firefighter between the ladder and the building extends the halyard.

Figure 4.30 Lower the ladder into the building.

BEAM RAISE

Step 1: When the desired location for the raise has been reached, the heeler places the ladder beam on the ground. The firefighter at the tip rests the beam on one shoulder while the heeler places one foot on the lower beam at the butt spur (Figure 4.31).

Figure 4.31 The firefighter at the butt places it on the ground.

Step 2: The heeler then grasps the upper beam with hands apart and the other foot extended back to act as a counterbalance (Figure 4.32).

Alternate method: An alternate method of heeling the ladder is to stand parallel to the ladder at the butt. Place one foot against the butt spur and position the other forward toward the tip of the ladder (Figure 4.33).

Step 3: The firefighter at the tip advances hand-over-hand down the beam toward the butt until the ladder is in a vertical position (Figure 4.34).

Figure 4.32 One method of heeling uses the foot closest to the tip.

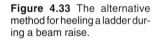

Figure 4.33 The alternative method for heeling a ladder during a beam raise.

Figure 4.34 Bring the ladder to vertical.

CAUTION: Visually check the area overhead for obstructions before bringing the ladder to a vertical position. Before stepping forward, visually check the terrain.

Step 4: Pivot the ladder to position the fly away from the building (fly in for wood ladders) if it is not already in that position. The firefighter on the halyard side of the ladder grasps the halyard and extends the fly section with a hand-over-hand motion (Figure 4.35). When the tip is at the desired elevation, the firefighters make sure the ladder locks are in place.

Step 5: To lower the ladder into position, the firefighter on the outside of the ladder places one foot either against a butt spur or on the bottom rung and grasps the rung or beams. Both firefighters gently lower the ladder into the building (Figure 4.36).

Figure 4.35 Extend the halyard.

Figure 4.36 Lower the ladder into the building.

Three-Firefighter Raise

As the length of the ladder increases, the weight also increases. This situation requires more personnel for raising the larger extension ladders. Typically, ladders of 35 feet (11 m) or larger should be raised by at least three firefighters. The procedure for raising ladders with three firefighters is as follows:

FLAT RAISE

Step 1: When the desired location for the raise has been reached, the heeler places the ladder butt on the ground while the firefighters at the tip rest the ladder flat on their shoulders (Figure 4.37).

Figure 4.37 Place the butt on the ground.

Step 2: The heeler heels the ladder either by standing on the bottom rung or by placing the toes or insteps on the beams. The heeler then crouches down to grasp a convenient rung with both hands, and leans back (Figures 4.38 a and b).

Figures 4.38 a and b One firefighter heels the ladder while the other two do the raising. These figures show the two methods of heeling a ground ladder.

CAUTION: Visually check the area overhead for obstructions before bringing the ladder to a vertical position. Before stepping forward, visually check the terrain.

Step 3: The firefighters at the tip advance in unison, with their outside hands on the beams and inside hands on the rungs, until the ladder is in a vertical position (Figure 4.39).

Figure 4.39 Raise the ladder keeping the outside hands on a beam and the inside hands on the rungs.

Step 4: If necessary, the firefighters pivot the ladder to position the fly section out from the building, using the procedure for three firefighters described earlier. If using a wood ladder, the fly should be in toward the building.

Step 5: To extend the ladder, one firefighter grasps the halyard and extends the fly section with a hand-over-hand motion while the other two steady the ladder (Figure 4.40). When the tip is at the desired elevation, the firefighters make sure that the ladder locks are in place.

Step 6: To lower the ladder into position, the firefighters on the outside of the ladder each place one foot either against a butt spur or on the bottom rung and grasp either the beam or a convenient rung.

Either method is acceptable as long as both do it the same way. The third firefighter steadies the ladder from the inside position. All firefighters gently lower the ladder into the building (Figure 4.41).

Figure 4.40 One firefighter extends the fly, while the others brace the ladder.

Figure 4.41 Lower the ladder into the building.

BEAM RAISE

To raise a ladder using the beam method with three firefighters, follow the same procedures for the two-firefighter raise. The only difference is that the third firefighter is positioned along the beam (Figure 4.42). Once the ladder has been raised to a vertical position, follow the procedures described for the flat raise.

Figure 4.42 One firefighter heels the ladder, while the other two perform the beam raise.

Four-Firefighter Flat Raise

When enough personnel are available, four firefighters can be used to better handle the larger and heavier extension ladders. A flat raise is normally used, and the procedures for raising the ladder are similar to the three-firefighter raise except for the placement of personnel. A firefighter at the butt is responsible for placing the butt at the desired distance from the building and determining whether the ladder will be raised parallel with or perpendicular to the building. The procedure for raising ladders with four firefighters is as follows:

Step 1: When the desired location for the raise has been reached, the heelers place the ladder butt on the ground while the firefighters at the tip rest the ladder flat on their shoulders (Figure 4.43).

Figure 4.43 Place the butt on the ground.

Step 2: The heelers heel the ladder by placing their inside feet on the bottom rung and their outside feet on the ground outside the beam. They each then grasp a convenient rung with the inside hand and the beam with the other hand and pull back (Figure 4.44).

CAUTION: Visually check the area overhead for obstructions before bringing the ladder to a vertical position. Before stepping forward, visually check the terrain.

Step 3: The firefighters at the tip advance in unison, with their outside hands on the beams and inside hands on the rungs, until the ladder is in a vertical position (Figure 4.45).

Figure 4.44 The heelers place their inside feet on the bottom rung.

Figure 4.46 Extend the fly section.

Figure 4.45 Bring the ladder to vertical.

Step 4: If necessary, pivot the ladder to position the fly section away from the building. Wood ladders should be positioned with the fly section in toward the building.

Step 5: To extend the ladder, either one or both of the firefighters on the halyard side of the ladder grasp the halyard and extend the fly section with a hand-over-hand motion (Figure 4.46). If both firefighters do this, they must coordinate their actions so as not to drop the fly section accidentally. When the tip is at the desired elevation, they make sure that the ladder locks are in place.

Step 6: To lower the ladder into position, the firefighters on the outside of the ladder place their inside feet against either the butt spur or bottom rung and grasp the beams (Figure 4.47). All firefighters gently lower the ladder into the building.

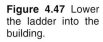

Figure 4.47 Lower the ladder into the building.

Pole Ladder Raises

Because of their length and weight, pole ladders require closely coordinated teamwork in order to be raised successfully and safely. A minimum of four firefighters are required to raise these ladders safely. Staypoles are used to assist in overcoming the problem of instability when the ladder is raised, extended, and lowered into the building. The staypoles provide a means for two additional firefighters to apply lifting force when the ladder is being raised and to take part of the weight when it is being lowered into position. When the ladder is vertical and the fly is being extended or retracted, the staypoles are used to provide both sideways and in-and-out stability.

When four firefighters raise a pole ladder, the operation must be performed perpendicular (at a right angle) to the building. Five- and six-firefighter raises may be performed either perpendicular or parallel to the building. The perpendicular operation is preferred because when the pole ladder is raised parallel to a building, a 90-degree pivot is necessary. This pivot requires that the staypoles be shifted at the same time the ladder is pivoted, which is somewhat complicated. The procedure is detailed at the beginning of the section on five-firefighter raises.

Pole ladders must be placed on the ground before beginning a raise. This procedure allows firefighters to flip the ladder over prior to raising, to shift from carrying to raising positions, and to pass the staypoles.

PASSING UNATTACHED STAYPOLES

Current NFPA requirements for pole ladder design dictate that staypoles be permanently attached to the beams of the ladder. However, many pole ladders built prior to the existence of this requirement have staypoles that are detachable. It is most common for these staypoles to be stored detached from the ladder, thus creating the need to attach them prior to raising the ladder (Figure 4.48). Unattached staypoles are nested between the beams and upon the rungs of the top fly section of the pole ladder. The toggle end of the staypole is carried so that it is nested at the butt of the ladder.

Unattached staypoles are passed differently than are attached staypoles. The procedures for raising pole ladders that follow this section are written for pole ladders with attached staypoles. Should your ladder have unattached staypoles, the following procedure should replace later directions for passing the staypoles:

Figure 4.48 The unattached staypoles are stowed on top of the pole ladder.

Step 1:   The firefighters who would normally pass attached staypoles proceed to the midpoint of the ladder and each grasps a staypole (Figure 4.49).

Figure 4.49 The firefighters prepare to pass the staypoles.

Step 2: The staypoles are lifted as the firefighters turn and walk toward the tip end of the ladder (Figure 4.50).

Step 3: The firefighters carry the poles forward until the toggles are opposite the toggle latches. Then, they shift to that point and latch the toggles (Figure 4.51).

Figure 4.50 Walk the staypoles into position.

Figure 4.51 Latch the staypoles into the holes on the beam of the ladder.

FOUR-FIREFIGHTER POLE LADDER RAISE

The following procedure is used for raising pole ladders with four firefighters:

Step 1: The ladder is placed flat on the ground with the butt spurs almost against the building (Figure 4.52). Two firefighters are at each end of the ladder.

Figure 4.52 Initially, the ladder is placed on the ground with the butt slightly away from the building and the fly up.

Step 2: If the ladder is intended to be positioned with the fly out, turn the ladder over so that the fly will be down. This step is not necessary if a fly-in ladder is used.

Step 3: Place the butt against the building (Figure 4.53).

Step 4: The firefighters at the tip move outward in order to receive the staypoles. They stand about 5 feet (1.5 m) apart. The two firefighters at the butt unlatch the spur end of the staypoles and lift the staypoles to begin the passing evolution (Figure 4.54).

Figure 4.53 The ladder is flipped so that the fly is down and the butt is placed against the building.

Figure 4.54 The two firefighters, who were at the butt, begin to swing the poles into position.

Step 5: The two firefighters with the staypoles walk toward the tip, while grasping the staypoles hand-over-hand as they move the staypoles toward vertical. They continually check overhead for obstructions (Figure 4.55).

Figure 4.55 The staypoles are rotated past vertical to get to the polemen.

Figure 4.58 The firefighters lift the tip of the ladder, rotate underneath, and each shoulders a beam of the ladder.

Step 6: The staypoles are brought to vertical and then lowered toward the two waiting firefighters (Figure 4.56).

Step 7: After the staypoles are passed, the two firefighters who passed them proceed to the tip. They kneel beside the beam facing the tip. Then, each grasps a convenient rung with the near hand (Figure 4.57).

Figure 4.56 Once the tips of the staypoles reach the waiting polemen, the firefighters who passed them can let go.

Step 9: The two firefighters at the tip walk toward the building, moving hand-over-hand down the beams (Figure 4.59). As soon as the ladder is high enough, the firefighters on the staypoles push forward to assist with raising the ladder (Figure 4.60).

Step 10: All firefighters continue raising the ladder until it is vertical against the building (Figure 4.61).

Step 11: The firefighters at the ladder butt grasp a low rung with their near hands, palms up. With their other hands, they grasp a rung about head high, palms down (Figure 4.62).

Step 12: The butt is shifted outward to the point where the ladder will be positioned for climbing (Figure 4.63).

Figure 4.57 The two firefighters kneel at the tip of the ladder, facing the polemen.

Step 8: The firefighters at the tip simultaneously lift the ladder tip and stand, using the leg muscles. They pivot under the ladder, grasp the nearest beam, and face the building (Figure 4.58).

Figure 4.59 The ladder is walked toward a vertical position.

Figure 4.60 The polemen help to raise the ladder by pushing on the staypoles.

Figure 4.63 The butt is pulled out to the approximate position that will be needed to afford a good climbing angle once it is extended.

Step 13: One of the firefighters on a staypole shifts toward the building until the staypoles now give four-way stability to the ladder. The two firefighters on the staypoles watch the tip of the ladder and control their respective forward and backward movements of the ladder (Figure 4.64).

Figure 4.61 The ladder is raised flush against the building.

Figure 4.62 The two firefighters grasp rungs in preparation for pulling out the butt of the ladder.

Figure 4.64 One poleman moves to a position parallel with the building.

Step 14: One of the two firefighters at the butt shifts to a position behind the ladder and faces the ladder. The other firefighter heels the ladder on the outside. The ladder is brought to vertical (Figure 4.65).

Figure 4.66 The fly is extended as necessary.

Figure 4.65 The ladder is brought back to vertical.

Step 15: The firefighter on the inside then raises the fly while the other firefighter helps to steady the ladder. The firefighter holding the staypole out in front of the ladder determines when the ladder is at the desired height (Figure 4.66).

Step 16: The ladder is then gently lowered into position (Figure 4.67).

Step 17: The firefighters handling the staypoles check for proper positioning and alignment of the ladder. They make adjustments as necessary.

Step 18: The firefighters handling the staypoles walk the staypoles toward the building, and then lower them to the ground (Figure 4.68).

CAUTION: The staypoles must not be wedged. They are not designed to carry the stresses put on the ladder. When set in this position, they are used only to prevent side slippage. If you cannot properly place one or both poles, then place neither.

Figure 4.67 Lower the ladder into position.

Figure 4.68 Set the staypoles into a position where they help support the ladder.

FIVE- AND SIX-FIREFIGHTER POLE LADDER RAISES

The procedure for raising a pole ladder with either five firefighters or six firefighters is essentially the same and so will only be described once. The only real difference is that when six firefighters are used, a second firefighter is placed at the butt of the ladder to assist with heeling the ladder and with pulling the halyard. The procedure is as follows:

Step 1: The ladder is lowered to the ground at the approximate raise location.

Step 2: If the ladder is intended to be positioned with the fly out, the ladder is turned over so that the fly will be down. This step is not necessary if a fly-in ladder is used.

Step 3: The firefighters take their positions for passing and receiving the staypoles. Two firefighters stand out from the tip, ready to receive the poles. Two stand at midpoint of the beams to pass the poles. The fifth (and sixth) stands at the butt, ready to heel the ladder (Figures 4.69 a and b).

Figure 4.69a The positions for five firefighters who are preparing to pass the staypoles.

Figure 4.69b When six firefighters are available, two firefighters position themselves at the butt of ladder.

Step 4: The firefighters at the beams pass the staypoles to the firefighters who moved out from the tip (Figure 4.70).

Figure 4.70 The staypoles are passed to the waiting firefighters.

Step 5: Simultaneously, the firefighters then take their positions for the raise. If there is one firefighter at the butt, he or she should crouch on the bottom rung and grasp a convenient rung with both hands. If there are two firefighters at the butt, they should each heel either the bottom rung or the beams with their inside foot and grasp a convenient rung (Figures 4.71 a and b).

Figure 4.71a When five firefighters are raising the ladder, one firefighter heels the ladder.

Figure 4.71b When six firefighters are raising the ladder, two firefighters heel the ladder.

The firefighters at the beams kneel beside the ladder, with their inside knees on the ground, and grasp a rung just below the toggles (Figure 4.72). The firefighters with the staypoles stand at the outside of the poles, one hand holding the spur end with the spur extending between the fingers, and the other hand holding the staypole at a comfortable distance up from its spur end. The staypoles should be as nearly in line with the beams as possible (Figure 4.73).

Figure 4.72 Two firefighters position themselves at the tip of the ladder.

Figure 4.73 The polemen should be in line with the beams of the ladder.

Step 6:   The firefighter at the butt gives the command to raise the ladder. The firefighters at the tip rise, bringing the ladder tip to shoulder level. They then pivot under the beams until they are facing the butt (Figure 4.74).

Figure 4.74 The firefighters at the tip of the ladder shoulder the ladder.

Step 7:   The firefighters at the tip raise their arms upward and walk hand-over-hand down the beams, raising the ladder. The heelman (heelmen) lean(s) back so that the body weight acts as a counterbalance to help the raise (Figure 4.75).

Step 8:   When the ladder is at an angle of about 45 degrees, the staypole firefighters assume most of the ladder's weight and push the ladder up. The firefighters walking the beams continue to assist with the raise by steadying the ladder as the firefighters on the poles do the actual raising (Figure 4.76).

Step 9:   When the ladder is vertical, one of the firefighters holding a staypole walks toward the building until the staypole is in line with the lateral plane of the ladder. The two staypoles now give four-way stability to the ladder. The two firefighters holding the staypoles watch the tip of the ladder and control their respective forward and backward movements of the ladder (Figure 4.77).

Figure 4.75 The ladder is raised toward vertical.

Figure 4.76 The polemen raise the ladder.

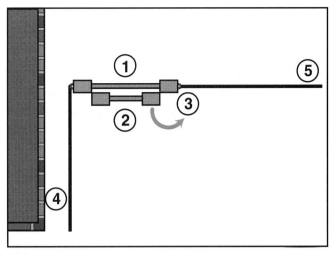

Figure 4.78 To begin the pivot, firefighter 3 shifts to the side away from the building. Firefighters 1 and 2 each place the near foot against the outside of the butt spurs.

Step 9b: Firefighters 1, 2, and 3 all work together to tilt the ladder slightly toward the building. They begin to pivot the ladder while continuing to steady it. Firefighter 5 shifts positions as the ladder pivots, keeping the staypole at 90 degrees to the building (Figure 4.79).

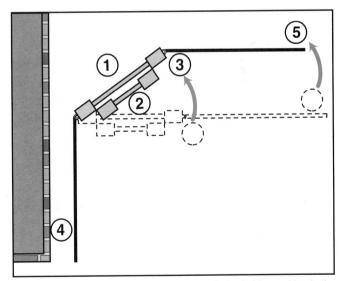

Figure 4.79 Firefighters 1, 2, and 3 all help tilt the ladder and begin the pivot. Firefighter 5 shifts as the ladder shifts.

Figure 4.77 When the ladder is vertical, one poleman moves to a parallel position.

NOTE: If the ladder has been raised parallel to the building, it is pivoted as described in Steps 9a, 9b, and 9c. (The numbers used to refer to firefighters are identified in the accompanying drawings.)

Step 9a: Firefighter 3 (one of the firefighters who walked the ladder to a vertical position) shifts to the side of the ladder away from the building. Firefighters 1 and 2 each place the near foot against the outside of the spur of the other beam (Figure 4.78).

Step 9c: Firefighters 1, 2, and 3 realign themselves with the ladder and continue the pivot until the ladder is at a right angle (perpendicular) to the building. Firefighter 5 continues to shift as the ladder pivots, being careful to maintain the staypole at 90 degrees to the building

(Figure 4.80). When six firefighters perform the pivot, the additional firefighter places both feet on the bottom rung, grasps a convenient rung with his or her hands, and rides the ladder around as it is pivoted. This firefighter's action places extra weight on the spur being pivoted to keep it from slipping.

Figure 4.81 Steady the ladder with the firefighters in position as shown.

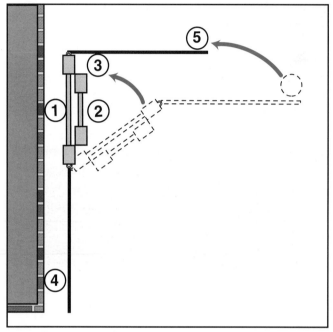

Figure 4.80 Firefighters 1, 2, and 3 reposition themselves and continue the pivot, bringing the ladder to right angles with the building.

Step 10: One of the firefighters who walked down the beams to raise the ladder shifts to the side of the beam and steadies the ladder from this position. The remaining firefighter on the outside of the ladder also steadies it. The firefighter on the side of the ladder next to the building grasps the halyard (Figure 4.81).

Step 11: The ladder is extended by pulling down on the halyard as previously described. When six firefighters are used, the two inside firefighters may pull the halyard together. The firefighter holding the staypole in front of the ladder determines when the ladder is at the desired height (Figure 4.82).

Step 12: The ladder is lowered into position (Figure 4.83).

Figure 4.82 Extend the fly section(s).

Figure 4.83 Lower the ladder into the building.

Figure 4.84 Set the staypoles.

Step 13: The firefighters handling the staypoles check for proper positioning and alignment of the ladder. They make adjustments as necessary.

Step 14: The firefighters handling the staypoles walk the staypoles toward the building, and then lower them to the ground (Figure 4.84).

CAUTION: The staypoles must not be wedged. They are not designed to carry the stresses put on the ladder. When set in this position, they are used only to prevent side slippage. If you cannot properly place one or both poles, then place neither.

SPECIAL PROCEDURES FOR RAISING SINGLE AND EXTENSION LADDERS

Sometimes the basic ladder raising procedures described will not in and of themselves be sufficient to get the ladder into its final position for use. In many cases, it will be necessary to move the ladder slightly after it has been extended. Often, obstructions will force firefighters to use alternative methods to raise the ladder. However, in all cases, it will be necessary to secure the ladder once it has been properly positioned. The following sections discuss some of these considerations.

Pivoting Ladders

Occasionally, a ladder will be raised with the fly in the incorrect position for deployment. When this happens, it will be necessary to pivot the ladder. Any ladder flat-raised parallel to the building will also require pivoting to align it with the wall upon which it will rest. The beam closest to the building should be used for the pivot. When pivoting a ladder, the procedures described in the following sections should be used.

ONE-FIREFIGHTER PIVOT

One-firefighter pivots should only be attempted with single ladders raised parallel to the building. This procedure is NOT RECOMMENDED for use with the one-firefighter extension ladder raise. The procedure described earlier in this chapter for the one-firefighter extension ladder raise details an alternative method of getting the fly section out. The procedure for the one-firefighter pivot is as follows:

Step 1: Place the foot nearest the building between the beams and against the butt spur of the beam nearest the building (Figure 4.85).

Step 2: Reach as low as convenient, and grasp the beam nearest the building. With your other hand, grasp the opposite beam as high as convenient (Figure 4.86).

Figure 4.85 The foot nearest the building is placed on the inside of the beam nearest the building.

Figure 4.86 Grasp the beam nearest the building low and the other one high.

Step 3: Tilt the ladder up on the beam just enough for the butt spur of the other beam to clear the ground.

Step 4: Simultaneously turn your body and shift the other foot as the ladder is pivoted 90 degrees (Figure 4.87).

Step 5: Bring the raised beam back down, and shift your position so that you are facing the ladder in a ready-to-climb position (Figure 4.88).

Figure 4.87 Make the pivot.

Figure 4.88 Lower the ladder into the building.

TWO-FIREFIGHTER PIVOT

The two-firefighter pivot may be used on any ground ladder that two firefighters can raise. The procedure described is for a ladder that must be turned 180 degrees to get the fly section in the proper position. The same procedure is used for positioning a ladder that was flat-raised parallel to the building. In this case, the beam nearest the building is used to pivot the ladder. The procedure is as follows:

Step 1: The two firefighters face each other through the ladder. They grasp the ladder with both hands, and one places a foot against the side of the beam on which the ladder will pivot (Figure 4.89).

Step 2: The ladder is then tilted onto the pivot beam (Figure 4.90).

Figure 4.89 Although both firefighters hold the ladder, only one heels a beam.

Figure 4.90 Tilt the ladder onto the pivot beam.

Step 3: The ladder is pivoted 90 degrees, with the firefighters simultaneously adjusting their positions (Figure 4.91).

Step 4: The process is repeated so that the ladder is turned a full 180 degrees and the fly is in the proper position.

Figure 4.91 Pivot the ladder 90 degrees.

Shifting Raised Ground Ladders

Occasionally, circumstances require that ground ladders be moved while vertical. Shifting a ladder that is in a vertical position should be limited to short distances, such as aligning ladders to a building or to an adjacent window.

ONE-FIREFIGHTER SHIFT

One firefighter can safely shift a single ladder that is 20 feet (6 m) long or less. The procedure is as follows:

Step 1: Face the ladder. Heel the ladder. Grasp the beams, and bring the ladder out to vertical (Figure 4.92).

Step 2: Shift your grip on the ladder, one hand at a time, so that one hand grasps as low a rung as convenient, palm upward. With your other hand, grasp a rung as high as convenient, palm downward (Figure 4.93).

Figure 4.92 Bring the ladder back to vertical.

Figure 4.93 The ladder is grasped as shown prior to making the shift.

Step 3: Turn slightly in the direction of travel. Visually check the terrain and the area overhead. Lift the ladder and proceed forward. Watch the tip as it is being moved (Figure 4.94).

> ## WARNING
> **Do not attempt this procedure close to live electrical wires.**

Figure 4.94 Watch the tip as the ladder is shifted.

Step 4: Set the ladder down at the new position. Once in place, switch your grip back to the beams. Heel the ladder, and lower it into position.

TWO-FIREFIGHTER SHIFT

Because of their weight, extension ladders require two firefighters for the shifting maneuver described as follows:

Step 1: If the ladder is not vertical, it is brought to vertical; and if extended, it is fully retracted (Figure 4.95).

Step 2: The firefighters position themselves on opposite sides of the ladder. The hands are positioned as for the one-firefighter shift except that the side grasped low by one firefighter is grasped high by the other (Figure 4.96).

Step 3: The ladder is then lifted just clear of the ground. While watching the tip, the firefighters shift to the new position (Figure 4.97).

Step 4: If necessary, the firefighters re-extend the ladder, then lower it gently into position.

Figure 4.95 The ladder must be fully retracted and vertical before starting the shift.

Figure 4.96 The firefighters prepare to make the shift.

Figure 4.97 Watch the tip as the ladder is shifted.

SHIFTING POLE LADDERS

Pole ladders may also be shifted vertically. At least four firefighters are required to safely perform this maneuver as follows:

Step 1: The ladder is brought back to vertical, and the fly(s) is (are) fully retracted.

Step 2: The two heelmen grasp the ladder the same as two firefighters moving an extension ladder. The firefighters holding the poles are in the normal position for extending and retracting (one parallel to the building and the other perpendicular to the building) (Figure 4.98).

Step 3: The heelmen lift the ladder (Figure 4.99). The polemen steady the ladder. The ladder is carried to the new position (Figure 4.100). When there are five or six firefighters, the additional firefighter(s) help steady the ladder.

Step 4: If necessary, the firefighters re-extend the ladder, then lower it gently into position, and redeploy the staypoles.

Figure 4.98 Two firefighters grasp the ladder while two others each take a staypole.

Figure 4.99 The ladder is lifted.

Figure 4.100 The polemen stabilize the ladder during the shift.

Rolling A Ladder

NFPA 1932 states that "ground ladders shall not be rolled beam over beam to reach a new position." Manpower at the time of need may not permit vertical shifting of an extension ladder, which is the alternative. Individual fire department policy should dictate whether or not rolling is permitted. IFSTA believes that it is necessary to make one possible exception to the NFPA standard. When one firefighter raises an extension ladder without assistance, the procedure, for reasons of safety, requires that the fly be in as it is extended. In order to get the fly out for climbing, a single roll is recommended rather than risking a 180-degree pivot with the fly extended. This procedure was detailed as a part of the description of the One-Firefighter Extension Ladder Raise earlier in this chapter.

Securing The Ladder

NFPA 1932 requires, and good common sense dictates, that ground ladders be secured whenever firefighters are climbing or working from them. The process of securing a ground ladder actually has as many as three steps:

Step 1: Make sure the pawls (dogs) are locked (extension ladders only).

Step 2: Tie the halyard (extension ladder only).

Step 3: Prevent movement of the ladder away from the building by heeling and/or tying in.

Obviously, the first two are not necessary when working with single ladders. Making sure the pawls are locked should have already been accomplished before the ladder was placed against the structure. The latter two are described in more detail in the following sections.

TYING THE HALYARD

Once the extension ladder is resting against the building and before it is climbed, the excess halyard should be tied to the ladder. This will prevent the fly from slipping or prevent anyone from tripping over the rope. The same tie can be used for either a closed- or an open-ended halyard. The procedure for tying the halyard is described as follows:

Step 1: Wrap the excess halyard around two convenient rungs. Pull it taut (Figure 4.101).

Step 2: Hold the halyard between your thumb and forefinger with your palm down (Figure 4.102).

Figure 4.101 Wrap the excess halyard around two rungs.

Figure 4.102 Hold the halyard between your thumb and forefinger.

Step 3: Turn your hand palm up and push the halyard underneath and back over the top of the rung (Figure 4.103).

Step 4: Grasp the halyard with your thumb and fingers and pull it through the loop, making a clove hitch (Figure 4.104). Finish the tie by making the half hitch or overhand safety on top of the clove hitch (Figure 4.105).

Figure 4.103 Push the halyard underneath and back over the rung.

Figure 4.104 Finish the clove hitch.

Figure 4.105 An overhand safety should be tied to secure the clove hitch.

HEELING

One method of preventing movement of the ladder is to properly heel, or foot, the ladder. There are several methods of properly heeling a ladder. One method is for the firefighter to stand underneath the ladder. The firefighter stands with feet about shoulder-width apart (or one foot slightly ahead of the other), grasps the ladder beams at about eye level, and pulls backward to press the ladder against the building (Figure 4.106). When using this method, the firefighter must wear head and eye protection and must not look up when there is someone climbing the ladder. The firefighter must be sure to grasp the beams and not the rungs.

Figure 4.106 The ladder may be heeled from behind.

Another method of heeling a ladder is for the firefighter to stand on the outside of the ladder and chock the butt with his or her foot (Figures 4.107 a and b). With this method, either the firefighter's toes are placed against the butt spur or one foot is placed on the bottom rung. The firefighter grasps the beams, and the ladder is pressed against the building. When a firefighter is heeling the ladder, he or she must stay alert for descending firefighters.

Figure 4.108a The ladder may be tied off near the bottom...

Figure 4.107a The ladder may be heeled with one foot on the rung.

Figure 4.108b ...or near the top.

Figure 4.107b The ladder may be heeled with a foot on the beam at the ground.

Raising Ladders Under An Obstruction

If hanging signs, overhead wires, or tree limbs prevent a normal ladder raise, it is still possible to raise the ladder under these obstructions. In these situations, a "tip-first raise" is required. The following is a method using two firefighters; however, three or four firefighters may be used if they are available:

Step 1: Place the ladder on the ground at 90 degrees to the building with the tip forward, approximately 3 feet (1 m) from the wall (Figure 4.109).

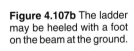
TYING IN

Whenever possible, a ladder should be tied securely to a fixed object. Tying in a ladder is simple, can be done quickly, and is strongly recommended to prevent the ladder from slipping or pulling away from the building. Tying in also frees personnel who would otherwise be holding the ladder in place. A rope hose tool or safety strap can be used between the ladder and a fixed object (Figures 4.108 a and b).

Figure 4.109 The ladder is placed on the ground with the tip about 3 feet (1 m) from the building.

Step 2:   One firefighter faces the building and kneels beside the ladder at the tip. The other firefighter heels the ladder (Figure 4.110).

Step 3:   The firefighter at the tip lifts it and pivots underneath, grasping a beam with each hand. When the pivot is complete, the arms are fully extended (Figure 4.111).

Step 4:   The heelman shifts to one side of the ladder, crouches, and grasps the second rung from the butt (Figure 4.112).

Step 5:   The firefighter at the tip remains stationary while passing the beams upward with the hands. The other firefighter walks forward, pulling the butt along the ground. A slight downward pressure is maintained to keep the butt from kicking up (Figure 4.113).

Step 6:   The firefighter at the butt stops pushing the ladder upward when it reaches the proper angle of inclination (Figure 4.114).

Step 7:   Set the ladder into the building (Figure 4.115).

Figure 4.112 The heelman shifts position to crouch beside the ladder and grasps the second rung.

Figure 4.113 The firefighter at the tip remains stationary while passing the beams upward with the hands. Momentum is provided by the other firefighter who walks forward pulling the butt along the ground.

Figure 4.114 The ladder is moved forward and up until the correct angle of inclination is attained.

Figure 4.110 The two firefighters are positioned as shown.

Figure 4.111 The tip is lifted. The firefighter pivots underneath and grasps a beam with each hand. The arms are fully extended.

Figure 4.115 When the ladder is at the proper spot, it is set against the building.

Dome (Auditorium) Raise

The dome raise, also called the auditorium raise, is a method of using a ground ladder to reach places where there is no means of supporting the tip of the ladder. Furthermore, the height needing to be reached requires a ladder longer than either an A-frame combination ladder or a single ladder supported by firefighters. This situation is found in some public buildings, churches, auditoriums, arenas, gymnasiums, skating rinks, or where high ceilings are required. Light fixtures suspended from high ceilings can also be reached with this raise.

Four guy lines are rigged from the tip. These lines are held by firefighters to provide the necessary support for the ladder. Two 125-foot (38 m) lifelines are used and a minimum of six firefighters are required — eight are preferred.

NOTE: All rope used for this evolution must meet the requirements for life safety rope contained in NFPA 1983, *Standard on Fire Service Life Safety Rope, Harness, and Hardware.*

The procedure for raising a ladder with six firefighters is described as follows:

Step 1: Place the ladder flat on the floor, fly up. Pull the top fly section out slightly to facilitate attaching the guy ropes (Figure 4.116).

Figure 4.116 The ladder lies on the ground with the fly slightly extended.

Step 2: String out the ropes, one from the tip of each beam, with the rope midpoint next to the tip of the ladder and the free ends away from it (Figure 4.117).

Figure 4.117 Two ropes are folded in half, and the bends (bights) are placed at the tip of the fly section.

Step 3: Pass the rope bend that is formed over the top of the beam, between the top two rungs, then under the beam to bring it back to the outside of the ladder. Then, pass the loop end over the rope to bring it toward the top (Figure 4.118). The loop is placed over the tip end of the beam (Figure 4.119).

Figure 4.118 The rope bend is wrapped around the beam and back over the rope itself.

Figure 4.119 The rope bends are slipped over the tips of the beams.

Step 4: Position a firefighter near the end of each guy line (Figure 4.120). Each passes the rope behind the body at the lower back or buttocks so that body weight is used to support the weight of the ladder on the guy lines (Figure 4.121).

Figure 4.120 The rope should be wrapped around each firefighter's back.

Figure 4.121 There will ultimately be four firefighters with ropes.

Step 5: The remaining two firefighters raise the ladder to vertical using a standard two-firefighter raise (Figure 4.122). If more than six firefighters are available, the additional firefighters assist these two firefighters with raising the ladder. The firefighters on the guy lines adjust their positions as the ladder is brought to vertical.

Figure 4.122 The ladder is raised to vertical.

NOTE: If a pole ladder is being used, the guy lines must be outside the poles before the ladder is raised to vertical.

Step 6: Extend the fly while the firefighters on the guy lines feed out sufficient rope. They shift position as necessary to maintain loose tension on the top of the ladder (Figure 4.123).

Step 7: When the ladder is extended to the desired height and the pawls have been engaged, the firefighters on the guy lines take up the slack on the ropes. They stand with their feet braced so that any sudden stresses put on the rope they are holding will not cause them to move suddenly and lose tension on the rope (Figure 4.124).

Step 8: Tie off the halyard prior to climbing the ladder.

Figure 4.123 Extend the fly section(s).

Figure 4.124 Once it is extended, the firefighters on the ropes stabilize the ladder.

RAISING OTHER LADDERS

Up to this point in the chapter our discussions have been limited to the raising of single and extension ladders. The remainder of this chapter is dedicated to the various types of special ladders used by firefighters.

Folding Ladder Raise

Folding ladders are most commonly used inside structures to gain access to concealed spaces such as attics or plenum spaces. They are usually carried to the place of deployment in a folded position and then unfolded immediately prior to being raised. The following procedure can be used:

Step 1: While folded, one beam projects further than the other. Place the foot pad of the projecting beam on the floor or ground (Figure 4.125).

Step 2: Open the ladder by pulling the beams apart until both beams rest firmly on the floor or ground and the rungs are level (Figure 4.126).

CAUTION: Care must be taken to keep from pinching the hands and fingers as this ladder is opened or closed.

Step 3: Lock the brace in place (Figure 4.127).

Step 4: Place the tip either against the wall or against the edge of the scuttle opening.

Figure 4.125 Place the lower beam on the ground.

Figure 4.126 Pull the beams apart to open the ladder.

Figure 4.127 Step on the brace to lock the beams.

Combination Ladder Raises

The method for raising a combination ladder depends on the exact type of combination ladder used. The following sections describe the deployment of each of the four major types of combination ladders.

COMBINATION EXTENSION/A-FRAME LADDER

This ladder may be used as either a short extension ladder or as an A-frame ladder. To erect this ladder as an extension ladder, the procedure is as follows:

Step 1: Grasp the ladder by both beams. Position the butt where the ladder is to be raised (Figure 4.128).

Figure 4.128 Place the butt where the ladder will be needed.

Step 2:   Shift one hand to grasp a rung on the fly section. Extend the fly by lifting upward (Figure 4.129).

Step 3:   Engage the pawls and lower the ladder tip into place.

Figure 4.129 Extend the fly.

To erect this ladder as an A-frame ladder, the procedure is as follows:

Step 1:   Lower the fly until the steel rods engage the slotted fittings at the top of the bed section (Figure 4.130).

Step 2:   Pull the lower part of the fly section outward away from its nested position inside the rails of the bed section (Figure 4.131).

Step 3:   When the tip end of the fly section is the proper distance from the butt of the bed section, a locking device at the top of the A-frame will prevent further spreading (Figure 4.132).

Figure 4.130 Lower the fly section until the rods are in the slots.

Figure 4.131 Pull the base sections apart.

Figure 4.132 A close-up of the locking device.

COMBINATION SINGLE/A-FRAME LADDER

A second type of combination ladder is a single ladder with each beam hinged at midpoint. It is usually racked and carried folded.

To erect this ladder as a single ladder, the procedure is as follows:

Step 1:   Flip up the top section until it latches to form a single ladder (Figure 4.133).

Figure 4.133 Swing the two sections into line to form a single ladder.

Step 2: Raise the ladder the same as any other single ladder.

To erect this ladder as an A-frame ladder, the procedure is as follows:

Step 1: Place the butt down. Shift the tip half out to form an A-frame (Figure 4.134).

Step 2: Lock the braces in place (Figure 4.135).

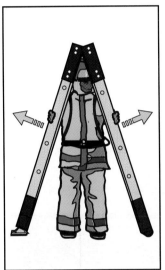

Figure 4.134 Pull the bases out.

Figure 4.135 Flip the locking braces into position to hold the ladder in place.

COMBINATION SINGLE/A-FRAME LADDER: TELESCOPING BEAM TYPE

A third type of combination ladder has two telescoping sections with the tops joined by a hinge. The rungs of the lower half of each section are mounted on the top of the beam rail so that they can slide inside the beam rail. It is carried folded with the telescoping sections retracted.

To erect this ladder as a single ladder, the procedure is as follows:

Step 1: Place the ladder on the ground, and rotate one side 180 degrees to form a single ladder (Figure 4.136).

Step 2: Lock the hinge in place, using the latching device that is incorporated into the hinge.

Step 3: If added height is needed, extend the telescoping beams within each section. Make sure these sections are properly locked after they have been extended (Figure 4.137).

Step 4: Raise the ladder the same as a single ladder.

To erect this ladder as an A-frame ladder, the procedure is as follows:

Step 1: Place the butts on the ground, and spread the two halves to attain the A-frame configuration. A latch in the hinge prevents further spreading (Figure 4.138).

Figure 4.136 This ladder may be used as a straight ladder.

Figure 4.137 The telescoping sections may be extended for greater reach.

Figure 4.138 This ladder makes a stable A-frame ladder.

Step 2: You may adjust the height of either or both sides (Figures 4.139 a and b). To do this, pull the pin on the beam outward, pull or push the section upward, and reinsert the pin when the desired height is attained.

Figure 4.139a This ladder may be used on even surfaces...

Figure 4.139b ... or on uneven surfaces.

EXTENDING A-FRAME LADDER

This ladder is not a true combination ladder because it is not normally used as an extension ladder; the tip is not placed against the wall. The extension is freestanding while supported by the A-frame.

To erect this ladder, the procedure is as follows:

Step 1: Place the ladder butts on the ground (Figure 4.140).

Step 2: Shift to face the ladder, spread the two A-frame sections, and set the brace latch (Figure 4.141).

Figure 4.140 Set the butt sections on the ground.

Locking device pressed downward to set

Figure 4.141 Spread the butt sections and lock the brace.

Step 3: Shift to the side, reach in between the A-frame, grasp the fly section, and push upward to extend it (Figure 4.142).

Step 4: When the desired height is reached, engage the pawls (Figure 4.143).

Figure 4.142 Extend the fly section.

Figure 4.143 Make sure the pawls are locked.

Pompier Ladder Raise

Pompier ladders are mainly used for training, but a few fire departments still use them in fireground operations. The following steps for raising pompier ladders are for drill tower use:

Step 1: Place the butt against the building.

Step 2: Raise the ladder to vertical by walking toward the building while gripping the beam hand-over-hand (Figure 4.144).

NOTE: The gooseneck hook is turned away from the building during this procedure.

Figure 4.144 Walk the pompier up into position.

Step 3: When the ladder reaches vertical, turn the gooseneck hook into the window opening, and set the hook over the windowsill (Figures 4.145 a and b).

Figure 4.145a Turn the hook into the window opening.

Figure 4.145b Set the hook over the windowsill.

Step 4: Test the security of the hook by pulling down on the ladder before beginning the climb (Figure 4.146).

The procedure for climbing the pompier ladder and moving it from floor to floor are covered in the next chapter.

Figure 4.146 Test the security of the hook on the windowsill.

Chapter 4 Review

Directions

The following activities are designed to help you comprehend and apply the information in Chapter 4 of **Fire Service Ground Ladders**, Ninth Edition. To receive the maximum learning experience from these activities, it is recommended that you use the following procedure:

1. Read the chapter, underlining or highlighting important terms, topics, and subject matter. Study the photographs and illustrations, and read the captions under each.

2. Review the list of vocabulary words to ensure that you know the chapter-related meaning of each. If you are unsure of the meaning of a vocabulary word, look the word up in the glossary or a dictionary, and then study its context in the chapter.

3. On a separate sheet of paper, complete all assigned or selected application and review activities before checking your answers.

4. After you have finished, check your answers against those on the pages referenced in parentheses.

5. Correct any incorrect answers, and review material that was answered incorrectly.

Vocabulary

Be sure that you know the chapter-related meanings of the following words.

- plenum *(139)*
- scuttle *(139)*

Application Of Knowledge

Choose from the following ladder raises those that are appropriate to your department and equipment, or ask your training officer to choose appropriate raises. Practice the chosen raises under your training officer's supervision.

- Perform a one-firefighter single ladder raise. *(110)*
- Perform a one-firefighter extension ladder raise (from the high-shoulder carry). *(111)*
- Perform a one-firefighter extension ladder raise (from the low-shoulder carry). *(113)*
- Perform a two-firefighter flat raise. *(114)*
- Perform a two-firefighter beam raise. *(115)*
- Perform a three-firefighter flat raise. *(117)*
- Perform a three-firefighter beam raise. *(118)*
- Perform a four-firefighter flat raise. *(118)*
- Perform a four-firefighter pole ladder raise (perpendicular to building). *(121)*
- Perform a five- and six-firefighter pole ladder raise (perpendicular or parallel to building). *(125)*
- Pass unattached pole ladder staypoles. *(120)*
- Perform a one-firefighter ground ladder pivot. *(130)*
- Perform a two-firefighter ground ladder pivot. *(130)*
- Perform a one-firefighter ground ladder shift. *(131)*
- Perform a two-firefighter ground ladder shift. *(132)*
- Perform a four-firefighter pole ladder shift. *(132)*
- Secure a raised single ladder. *(133)*
- Secure a raised extension ladder. *(133)*
- Tie an extension ladder halyard. *(134)*
- Heel a ladder. *(134, 135)*
- Tie in a ground ladder. *(135)*
- Perform a two-firefighter ladder raise under an obstruction. *(135)*
- Perform a dome (auditorium) raise. *(137)*
- Raise a folding ladder. *(139)*
- Raise a combination extension/A-frame ladder. *(139)*
- Raise a combination single/A-frame ladder. *(140)*
- Raise a combination single/telescoping beam A-frame ladder. *(141)*
- Raise an extending A-frame ladder. *(142)*
- Raise a pompier ladder. *(143)*

Review Activities

1. Describe things to consider about electrical hazards before raising a ladder. *(107)*

2. Discuss the pros and cons of fly position on extension ladders. Be sure to include general rule of thumb guidelines. *(108)*

3. Explain how the firefighter determines whether to perform a parallel or a perpendicular ladder raise. *(109)*

4. Explain the major difference between the one-firefighter extension ladder raise from the low-shoulder carry and the same raise from the high-shoulder carry. *(111, 113)*

5. Answer the following questions:
   - As ladder length increases, the ladder's weight also increases. Typically, at what ladder length should at least three firefighters be used to raise the ladder? *(117)*
   - What is the minimum number of firefighters required to raise a pole ladder safely? *(120)*
   - When four firefighters raise a pole ladder, how must the operation be performed in regard to ladder orientation to the building? *(120)*
   - What do current NFPA requirements for pole ladder design dictate about staypoles? *(120)*
   - Where are unattached staypoles stored? *(120)*

6. Explain why ladders must sometimes be pivoted. *(129)*

7. Discuss the NFPA and IFSTA positions on rolling a ladder. *(133)*

8. Explain what a dome raise is and when it might be necessary. *(137)*

Questions And Notes ---

Chapter 5

Climbing And Using
Ground Ladders

LEARNING OBJECTIVES

This chapter provides information that will assist the reader in meeting the objectives contained in NFPA 1001, *Standard for Fire Fighter Professional Qualifications* (1992 edition). The objectives contained in this chapter are as follows:

Fire Fighter I

3-11.3 Demonstrate the procedures of working from ground or aerial ladders with tools and appliances, with and without a safety harness.

3-11.4 Climb the full length of each type of ground and aerial ladder available to the authority having jurisdiction and demonstrate:

    (a)   Carrying fire fighting tools or equipment while ascending and descending

    (b)   Bringing an injured person down the ladders

3-11.5 Demonstrate the deployment of a roof ladder on a pitched roof.

3-12.5 Demonstrate operation of a charged attack line 1 1/2-in. (38 mm) or larger from a ground ladder.

Fire Fighter II

4-18.4 Raise and lower a person a maximum of 20 vertical ft (6 m) with a rope rescue system.

Chapter 5
Climbing And Using Ground Ladders

The information provided to this point in the manual has instructed the firefighter on how to select an appropriate ground ladder, how to carry the ladder to the point of deployment, and how to efficiently raise the ladder. Once these tasks have been completed, the ladder is ready for the use for which it was intended. The firefighter can now climb the ladder, perform the required task, and then descend to the point of deployment (usually the ground). This chapter contains information on how to safely climb and work from ladders, as well as information related to specific tactical uses for ladders.

CLIMBING LADDERS

Ladder climbing should be done smoothly and rhythmically. The climber should ascend the ladder so that there is the least possible amount of bounce and sway. This smoothness will be accomplished if the climber's knee is bent to ease the weight on each rung. Balance on the ladder will come naturally if the ladder is properly spaced from the building, for the body will be perpendicular to the ground.

The climb may be started after the climbing angle has been checked and the ladder is properly secured. The climber's eyes should be focused forward, with an occasional glance at the tip of the ladder. The climber's arms should be kept straight during the climb; this action keeps the body away from the ladder and permits free knee movement during the climb (Figure 5.1). The hands should grasp the rungs with the palms down and the thumbs beneath the rung. Some people find it natural to grasp every rung with alternate hands while climbing; others prefer to grasp alternate

rungs (Figure 5.2). An option for hand placement when climbing ground ladders is to climb while sliding both hands up behind the beams to maintain constant contact, as when climbing and carrying equipment (Figure 5.3). Firefighters should try both methods and select the one that is either most natural or is required by departmental SOP.

Figure 5.1 Proper ladder climbing techniques include keeping both the back and arms straight as the climb is made.

Figure 5.2 Some firefighters find that there is less "bounce" to the climb if the foot and hand on the same side are raised together.

Figure 5.3 The hands may be slid up the underside of the beams.

Figure 5.4 Slide one hand up the underside of the beam, and carry the tool in the other hand.

Figure 5.5 Some firefighters prefer to slide the hand carrying the tool up the beam, as well as the free hand.

If the feet should slip, the arms and hands are in a position to stop the fall. All upward progress should be performed by the leg muscles, not the arm muscles. The arms and hands should not reach upward during the climb, because reaching upward will bring the body too close to the ladder.

Practice climbing should be done slowly to develop form rather than speed. Speed will be developed as the proper technique is mastered. Too much speed results in lack of body control, and quick movements cause the ladder to bounce and sway.

Often during fire fighting, a firefighter is required to carry equipment up and down a ladder. This procedure interrupts the natural climb either because of the added weight on the shoulder or because of the necessity of using one hand to hold the tool. If the tool is to be carried in one hand, it is desirable to slide the free hand under the beam while making the climb (Figure 5.4). This method permits constant contact with the ladder. If the firefighter's hands are large enough, the hand with the tool may also be slid along the beam (Figure 5.5). Whenever possible, a handline rope should be used to hoist tools and equipment rather than carrying them up the ladder.

The technique used for climbing aerial ladders is basically the same as that for ground ladders. The only difference is that the firefighter has the option of grasping the handrails as well as the rungs when climbing (Figure 5.6).

Working On A Ladder

Firefighters must sometimes work while standing on a ground ladder, and both hands must be free. Either a Class I life safety harness (ladder belt) or a leg lock can be used to safely secure the firefighter to the ladder while work is being performed.

Figure 5.6 Use the hand rails when climbing an aerial ladder.

WARNING
Never use a leg lock on an aerial ladder.

The life safety harness must be strapped tightly around the waist during use. The hook may be moved to one side, out of the way, while the firefighter is climbing the ladder. However, after reaching the desired height, the firefighter returns the hook to the center and attaches it to a rung (Figure 5.7). All life safety harnesses should meet the requirements set forth in NFPA 1983, *Standard on Fire Service Life Safety Rope, Harness, and Hardware.*

Figure 5.7 One method of securing to a ladder is to use a ladder or safety belt.

The following steps should be used when applying a leg lock on a ground ladder:

Step 1: Climb to the desired height.

Step 2: Advance one rung higher (Figure 5.8).

Step 3: Slide the leg on the opposite side from the working side over and behind the rung that you will lock onto (Figure 5.9).

Step 4: Hook your foot either on the rung or on the beam (Figures 5.10 a and b).

Step 5: Rest on your thigh.

Step 6: Step down with the opposite leg (Figure 5.11).

Figure 5.8 Climb to a level that is one rung higher than the one you will stand on.

Figure 5.9 The leg on the side opposite the working side is slid over the rung.

Figure 5.10a The foot may be hooked over a rung...

Figure 5.10b ...or around the near beam.

Figure 5.11 Step down with the leg that is not locked.

Placing A Roof Ladder

There are a number of ways to get a roof ladder in place on a sloped roof, using either one or two firefighters. These procedures are highlighted below.

ONE-FIREFIGHTER ROOF LADDER DEPLOYMENT

One firefighter can conveniently carry a roof ladder to the roof by using the shoulder method. The hooks should be closed while the ladder is being carried. The following procedure shows one method of placing a roof ladder in position:

Step 1: Carry the roof ladder to the ladder that is to be climbed. Set the roof ladder down and open the hooks (Figure 5.12).

Step 2: With the hooks facing outward, tilt the roof ladder up so that it rests against the other ladder (Figure 5.13).

Step 3: Climb the main ladder until your shoulder is about two rungs above the midpoint of the roof ladder (Figure 5.14).

Step 4: Reach through the rungs of the roof ladder and hoist it onto your shoulder (Figure 5.15).

Step 5: Climb to the top of the ladder, and use a leg lock or life safety harness to lock into the ladder (Figure 5.16).

Step 6: Once locked in, take the roof ladder off your shoulder, and use a hand-over-hand method to push the roof ladder onto the roof (Figure 5.17). The ladder should be pushed onto the roof so that the hooks are in the down position.

Step 7: Push the roof ladder up the roof until the hooks go over the edge of the peak and catch solidly (Figure 5.18).

Step 8: Remove the roof ladder by reversing the process.

Figure 5.12 Open the hooks of the roof ladder at the base of the extension ladder.

Figure 5.13 The hooks should be facing outward as the roof ladder is tilted against the extension ladder.

Figure 5.14 The roof ladder should be shouldered about two rungs above midpoint.

Figure 5.15 The roof ladder is hoisted onto the firefighter's shoulder.

Figure 5.16 Prepare to lock or hook into the extension ladder.

Figure 5.17 The hooks should be facing downward as the roof ladder is pushed onto the roof.

Figure 5.18 The hooks should be firmly over the peak of the roof.

TWO-FIREFIGHTER ROOF LADDER DEPLOYMENT

It is much easier to climb another ladder and place the roof ladder using two firefighters. There are two methods of accomplishing this task, both named for the way the ladder is carried from the apparatus: Hooks-First Method and Butt-First Method. The Modified Butt-First Method will also be discussed.

Hooks-First Method

Step 1:   Carry the ladder using the low-shoulder method, hooks (tip) first. The firefighter at the tip opens the hooks in such a manner that the hooks face outward (Figure 5.19).

Figure 5.19 The firefighter at the tip of the ladder opens the hooks.

Step 2: The two firefighters ascend the other ladder, using their free hands on the beam for support (Figure 5.20). On short raises, it may not be necessary for the bottom firefighter to ascend the ladder.

Step 3: When the top firefighter reaches the roof edge, he or she either leg locks in or connects his or her safety belt to the ladder (Figure 5.21).

Step 4: Both firefighters remove the roof ladder from their shoulders and push it on its beam up onto the roof (Figure 5.22).

Figure 5.20 The firefighter at the tip begins to climb the extension ladder.

Figure 5.21 The firefighter at the tip locks or hooks into the extension ladder.

Figure 5.22 The ladder is pushed onto the roof on one beam.

Step 5: Slide the roof ladder up the roof on its beam until the balance point is reached. Then turn it flat, hooks down, and slide it the remaining distance to the roof peak on the hooks (Figure 5.23).

Step 6: After the hooks drop over the peak, the firefighter pulls back on the roof ladder to snug it in.

Figure 5.24 Set the butt of the roof ladder at the base of the extension ladder.

Figure 5.25 Raise the roof ladder to vertical.

Figure 5.23 Flip the ladder to a flat position with the hooks facing downward.

Butt-First Method

When a roof ladder has been carried to the scene butt first, there is no need to waste valuable time turning it around.

NOTE: The following procedure is intended for use with one- or one-and-one-half story buildings where the eaves are less than 14 feet (4 m) off the ground. For multiple story buildings, the modified butt first method may be used.

Step 1:   Lower the butt of the roof ladder to the ground adjacent to the ladder that has been raised to the roof. The firefighter at the tip maintains the carry position (Figure 5.24).

Step 2:   The firefighter who carries the butt assumes a beam raise heelman position. The firefighter at the tip shifts out of the carry position and raises the ladder to vertical. Lay the roof ladder on its beam edge alongside the other ladder. One of the firefighters steadies it (Figure 5.25).

Step 3:   One firefighter heels the climbing ladder while holding the roof ladder. The other firefighter climbs to a point near the tip, leg locks in, and opens the hooks away from the body (Figure 5.26).

NOTE: Step 3 may be omitted if it is possible to either secure the butt of the other ladder with a rope or use a third firefighter as the heelman.

Figure 5.26 The firefighter on the extension ladder opens the hooks on the roof ladder.

Step 4: Both firefighters grasp a convenient rung of the roof ladder, and push it upward. The firefighter at the tip grasps the roof ladder and slides it up the roof on one beam (Figure 5.27).

Figure 5.27 Slide the roof ladder onto the roof.

Step 5: Slide the roof ladder up the roof on its beam until the balance point is reached. Then turn it flat, hooks down, and slide it the remaining distance to the roof peak on its hooks (Figure 5.28).

Step 6: After the hooks drop over the peak, pull back on the roof ladder to snug it in.

Figure 5.28 Flip the ladder to a flat position with the hooks facing downward.

Modified Butt-First Method

Step 1: Lower the butt of the roof ladder to the ground approximately 1 foot (0.3 m) from the side of the other ladder and approximately 1 foot (0.3 m) closer to the building than the butt of the other ladder (Figure 5.29).

Step 2: The firefighter who carries the butt of the roof ladder becomes the heelman for raising it. Raise the roof ladder to vertical (Figure 5.30). If a flat raise is used, the ladder is then pivoted. Lay the roof ladder against the outside beam of the other ladder (Figure 5.31).

NOTE: It is important that the roof ladder be placed so that it extends several rungs above where it rests on the beam of the other ladder.

Step 3: The firefighter who heels the roof ladder as it is raised shifts position and heels the climbing ladder.

NOTE: Step 3 may be omitted if it is possible to either secure the butt of the other ladder with a rope or to use a third firefighter as a heelman.

Step 4: The uncommitted firefighter climbs the ladder until parallel to a point two rungs above the midpoint of the roof ladder. This firefighter extends the near arm between the rungs of the roof ladder and places the upper rung on the shoulder (Figure 5.32).

Figure 5.31 Rest the roof ladder against the extension ladder.

Figure 5.32 The roof ladder should be shouldered about two rungs above midpoint.

Step 5: The firefighter then continues climbing to the eaves. The heelman helps by feeding the roof ladder up as far as possible.

Step 6: The firefighter climbing the ground ladder either leg locks in or uses a safety belt at the top of the ground ladder (Figure 5.33).

Figure 5.29 Place the butt of the roof ladder about one foot (0.3 m) away from the base of the ground ladder and one foot (0.3 m) closer to the building than the base of the ground ladder.

Figure 5.30 Raise the roof ladder to vertical.

Figure 5.33 The firefighter on the ground helps hold the roof ladder as the firefighter at the tip locks or hooks in.

Step 7: Slide the roof ladder up the roof on the beam until the balance point is reached. Then turn it flat, hooks down, and slide it the remaining distance to the roof peak on its hooks (Figure 5.34).

Step 8: After the hooks drop over the peak, pull back on the roof ladder to snug it in.

Figure 5.34 Slide the roof ladder onto the roof.

Climbing A Pompier Ladder

The climbing procedures described for a pompier ladder are for drill tower use. The main difference between drill tower use and fireground use is that the windows in drill towers have been removed. In real fire situations, windows may be either entirely or partially intact. It may be necessary to break through the screen and to break out glass sufficiently to let the gooseneck hook through. When climbing, it would be necessary to hook in at the top of the pompier ladder and then clear the window opening with some type of forcible entry tool.

The section on raising ladders dealt with getting the pompier ladder in place initially. The following steps detail correct climbing procedures:

Step 1: Grasp the beam with both hands, and place your feet on the rungs close to the beam (Figure 5.35).

Step 2: Grasping the gooseneck hook, step into the window opening and straddle the windowsill (Figure 5.36).

Step 3: While astride the windowsill, dislodge the gooseneck, and turn it outward (Figure 5.37).

Figure 5.35 Keep your back straight and avoid a bouncing motion as the pompier ladder is climbed.

Figure 5.36 Grasp the gooseneck of the pompier ladder while straddling the windowsill.

Figure 5.37 The gooseneck is turned outward after it is lifted from the windowsill.

Step 4: Grasp the beam and raise the ladder vertically hand-over-hand until the gooseneck hook is opposite the next window up (Figure 5.38).

Step 5: Turn the gooseneck hook into the window opening, and secure it over the windowsill (Figure 5.39).

Step 6: Stand on the windowsill, grasp the beam, and step out onto the pompier ladder to climb to the next floor (Figure 5.40).

Figure 5.38 Use a hand-over-hand motion to raise the pompier ladder to the next window.

Figure 5.39 Turn the hook into the next window.

Figure 5.40 Step from the windowsill onto the pompier ladder.

TACTICAL OPERATIONS USING GROUND LADDERS

The entire purpose behind using ground ladders is to achieve some type of tactical purpose. These tactical purposes include things such as rescue, ventilation, and fire attack. The remainder of this chapter is dedicated to providing information pertaining to using ladders for these specific tactical objectives.

Rescue Operations

Rescue is the first priority on the fireground, and ground ladders are one means of rescuing trapped victims from upper floors of the fire building. However, using ground ladders or aerial devices should be a last resort for removing these victims (Figure 5.41). When an interior stairway is available, it should be used for evacuation. If it is necessary to use ground ladders to rescue victims, certain tactics can be used to achieve faster, safer, and more efficient rescue operations.

Figure 5.41 Rescues may be made from ground or aerial ladders. *Courtesy of Bill Tompkins.*

PRIORITY CONSIDERATIONS

In situations that require using ground ladders for rescue, the main objective is to reach as many victims or points of egress as possible with a minimum amount of ladder maneuvering. Victims should be removed in the following order of priority:

1. People most severely threatened by current fire conditions

2. Largest number or groups of people

3. Remainder of people in the fire area

4. People in exposed area

It is obvious that those who are in the greatest amount of danger should be given the highest priority. Determining which victims are in the most danger is a judgment call that must be made by the company officer or incident commander. Typically, those occupants located either on or immediately above the fire floor will be in greatest danger (Figure 5.42). Visible fire conditions will be a strong indication of which victims are in the worst situation. Additional preference should be given to those individuals who are in a panicked state and may attempt to jump if they do not see help arriving soon.

Figure 5.42 Occupants on or above the fire floor receive the highest priority.

The second priority to be considered involves multiple victims who may be located in different parts of the fire building. When two or more groups of victims appear to be in the same amount of danger, the larger of the two groups should have

ground ladders extended to them first (Figure 5.43). In a worst case scenario, it is best to rescue the largest number of people possible, given the time available.

Figure 5.43 When only one ladder is available, raise it first to the location with the most victims.

The third priority involves the remainder of people in the fire area. Remaining groups of victims should be removed in descending order of numbers. During this process, the firefighters should continue to monitor fire conditions for changes that might increase the danger presented to any of the waiting victims.

The fourth priority is the people in the exposed area. Once all those who appear to be in imminent danger have been removed, other victims who are unable to escape the fire building or exposures by standard means may be evacuated.

ASSISTING CONSCIOUS AND PHYSICALLY ABLE PEOPLE

Evacuation of people from a building by ground ladders is difficult and slow. As stated previously, it is better to evacuate down stairways, through connecting buildings, or over connecting roofs. People being evacuated are generally unaccustomed to climbing down a ladder, and they must be accompanied by firefighters to prevent injury while descending. If a person panics and becomes violent during this operation, the recommended procedure is for the firefighter to pin the person to the ladder until he or she calms down.

The firefighters' actions during evacuation of conscious, physically able people are as follows:

Step 1: If the situation and personnel permit, one or more firefighters ascend the ladder and enter the building. Another firefighter climbs to the top of the ladder. All assist the victim onto the ladder (Figure 5.44).

Figure 5.44 One firefighter should be in the building to assist the victim onto the ladder and another firefighter should be on the ladder.

Step 2: As soon as the victim is on the ladder, the firefighter on the ladder places his or her arms around the victim and under the victim's armpits. The firefighter's hands grasp the ladder rung in front of the victim's face. One knee is placed between the victim's legs to provide support in case a rung is missed or unconsciousness occurs (Figure 5.45).

NOTE: The firefighter explains what is occurring and reassures the victim.

Step 3: The victim and the firefighter descend the ladder together (Figure 5.46).

Figure 5.45 The firefighter on the ladder should keep his or her arms beneath the victim's armpits.

Figure 5.46 Descend the ladder at a speed the victim is comfortable with.

BRINGING UNCONSCIOUS PEOPLE DOWN GROUND LADDERS

Assuming that the victim has been located and moved to the window area, a minimum of two firefighters are required to get the victim down the ladder. The following procedure is used:

Step 1: A firefighter climbs to near the top of the ladder. He or she places one foot on a rung in such a way that the knee is bent 90 degrees and the upper leg is in a horizontal plane (Figure 5.47).

Figure 5.47 The firefighter on the ladder should have one thigh in a horizontal position, ready to support the victim.

Step 2: The firefighter(s) inside the building strap the victim's hands together, and pass the victim out the window in such a manner that the victim is facing the firefighter on the ladder. The victim's arms are around the firefighter's neck (Figure 5.48).

Step 3: The firefighter on the ladder reaches under the victim's armpits and grasps a rung behind the victim's head (Figure 5.49).

Step 4: The firefighter holds the victim to the ladder with his or her arms under the armpits of the victim. The firefighter then positions a knee one rung lower. The victim is then let down onto the firefighter's knee, and the firefighter readjusts his or her grasp one rung lower (Figure 5.50).

Step 5: Repeat this procedure one rung at a time down the ladder.

Step 6: When the ground is reached, grasp the victim under the armpits and pull clear of the danger area. If there is another firefighter, he or she helps carry the victim to a safe location (Figure 5.51).

Figure 5.49 The rescuer grasps the rung behind the victim's neck.

Figure 5.48 The victim is passed to the firefighter on the ladder.

Figure 5.50 The ladder is descended one rung at a time.

Figure 5.51 Another firefighter assists the rescuer in removing the victim at the bottom of the ladder.

An alternate method of rescuing an unconscious victim is to have the victim facing the firefighter, but with the victim's knees over the firefighter's shoulders (Figure 5.52). This method works best if the ladder is placed at an angle slightly steeper than normal. The victim's armpits are supported by the firefighter's forearms. The firefighter descends the ladder rung by rung, while sliding his or her hands down the beams. In order to increase the control over the victim, the firefighter can lean in toward the ladder to slow the process.

Smaller-sized adults and children can be brought down a ladder by cradling them across the firefighter's arms (Figure 5.53).

Figure 5.52 An alternative method is to place the victim's legs over the rescuer's shoulders.

Figure 5.53 The firefighter cradles the infant while descending the ladder.

REMOVING AN UNCONSCIOUS FIREFIGHTER LEG LOCKED ON A GROUND LADDER

A special rescue procedure is required to remove a firefighter who becomes unconscious while leg locked onto a ladder. The steps for performing this procedure are as follows:

Step 1: One firefighter climbs the original ladder and pins the unconscious firefighter against the ladder by placing his or her hands under the unconscious firefighter's armpits, grasping a rung, and then leaning against the victim (Figure 5.54).

Figure 5.54 Pin the unconscious firefighter to the ladder.

Step 2: A second ladder is placed and raised immediately adjacent to the one containing the unconscious firefighter. It is placed on the same side as the unconscious firefighter's leg lock (Figure 5.55).

NOTE: If personnel are available, a third ladder is raised on the other side of the original ladder. The third ladder will be necessary if the unconscious firefighter is a large person.

Step 3: A second firefighter climbs the second ladder to the same level as the unconscious firefighter's foot on the locked leg. He or she frees the unconscious firefighter's foot from the rung, pushes it back through the space between the two rungs so that it is hanging free, and removes the boot (Figure 5.56).

Step 4: The firefighter on the second ladder (and the one on the third ladder if used) now climb to a level slightly above the unconscious firefighter.

Step 5: Working as a team, all firefighters push/ slide the unconscious firefighter up the ladder until the leg that was locked is removed from the rungs (Figure 5.57).

Step 6: Use the same method as described for removing an unconscious adult victim to bring the unconscious firefighter down the ladder. The only difference is that in all likelihood, the unconscious firefighter will be facing away from the rescuer (Figure 5.58).

When speed is a factor in freeing the unconscious firefighter, or when the firefighter is a very large person, and the rescuers have been unable to remove the leg from the leg lock, the alternative is to cut the rung that was used in the leg lock.

The normal position is taken to pin the unconscious firefighter to the ladder. The second ladder is raised next to the one holding the unconscious firefighter as explained in the previous procedure. The firefighter on the second ladder clears the unconscious firefighter's foot, but instead of trying to raise the firefighter to clear the leg, the rung it is over is cut away. The cutting is done by the firefighter

on the second ladder (Figure 5.59). When the rung is removed, the leg is pulled through, and the previously described procedure for removing an unconscious firefighter is used.

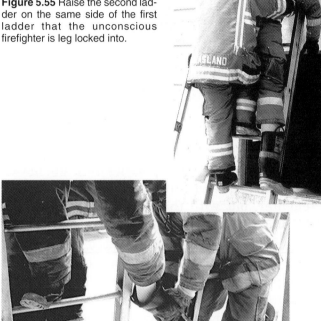

Figure 5.55 Raise the second ladder on the same side of the first ladder that the unconscious firefighter is leg locked into.

Figure 5.56 The firefighter on the second ladder frees the unconscious firefighter's leg and removes the boot.

Figure 5.57 Push the unconscious firefighter up the ladder to remove the locked leg from the rung.

Figure 5.58 The rescuer uses one thigh to support the unconscious firefighter while descending the ladder.

Figure 5.59 It may be necessary to cut the rung that the unconscious firefighter is locked into.

LOWERING EXTENSION LADDERS TO BELOW-GRADE LOCATIONS

Sometimes a ladder is required for access to victims in a below-grade location. An extension ladder may be required. If that is the case, the ladder is lowered after having been extended. The steps are as follows:

Step 1: Lay the ladder flat on the ground. Extend the fly by having firefighters anchor the bed section while others grasp a rung of the fly section and pull it outward (Figure 5.60).

Step 2: When the desired length is attained, latch the pawls, and tie the rungs of the ladder sections together (Figure 5.61).

Step 3: Tie a rope to the tip of the bed section. Tilt the ladder up on one beam. String the rope along the ground where the ladder was previously resting (Figure 5.62).

Figure 5.61 Tie the rungs together to prevent the ladder from extending or retracting.

Figure 5.62 Tie a rope to the tip of the ladder.

Figure 5.60 Pull the fly out while the ladder is lying on the ground.

Step 4: Lay the ladder back flat. The rope is now under the ladder. String the remaining rope out from the tip (Figure 5.63).

Step 5: At least two firefighters grasp the free end of the rope and take up the slack. Pick up the ladder and carry it, butt first, to the edge of the precipice. The firefighters holding the rope follow (Figure 5.64).

Step 6: Place the butt on the ground at the edge of the precipice, and shove the ladder outward until the balance point is reached. Slack off enough rope to allow the ladder to move forward. Allow the ladder to swing downward, at which time the firefighters on the rope take the weight of the ladder. The firefighters who were carrying the ladder now steady it against sideways movement (Figure 5.65).

Step 7: Just before the ladder butt reaches the ground, the firefighters steadying the ladder pull the top toward them to cause the butt to kick outward enough to obtain a better angle for climbing (Figure 5.66).

NOTE: It may be necessary to reset the butt when the first firefighter reaches the bottom.

Figure 5.65 Using the rope, lower the ladder over the edge.

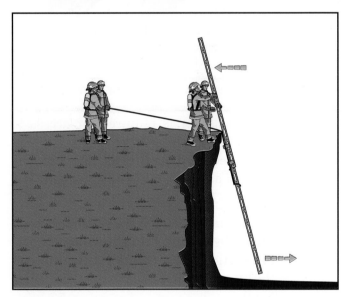

Figure 5.66 Kick the bottom of the ladder outward to produce a proper climbing angle.

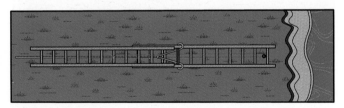

Figure 5.63 The rope is beneath the ladder.

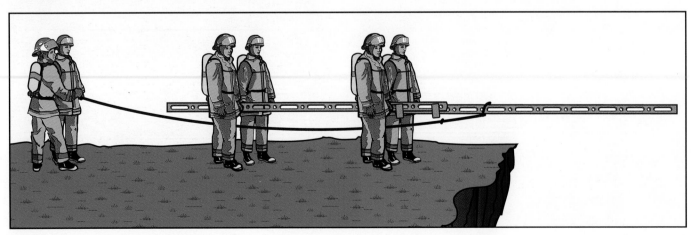

Figure 5.64 Carry the ladder to the edge of the cliff.

USING A LADDER TO SUPPORT LOWERING OF AN INJURED PERSON (LADDER SLING)

This evolution may be used when either unconscious or conscious victims have to be lowered from an upper floor. Although it may appear slow compared to carrying the victim down a ladder, this method may be safer under certain conditions. If several people are to be rescued, it may also prove to be more rapid than originally expected.

Step 1:   Raise a ladder to a point just above the window from which the rescue is to be made. The ladder is positioned in the same manner as if a fire stream were going to be operated from it.

Step 2:   If a double loop figure of eight is used, the knot may be either tied on the ground before running the rope up the ladder or tied in the building after running the rope up the ladder.

Step 3:   One firefighter takes the end of the lifeline, passes it underneath the bottom rung to the underside of the ladder, and starts the climb (Figure 5.67). Hold the rope in either hand so that it feeds up the underside of the ladder.

Step 4:   When the windowsill is reached, thread the rope back through the rungs toward the firefighter on the ladder, up and over three or four consecutive rungs (Figure 5.68). This procedure allows the rope to hang freely, but not too close to the building.

Step 5:   Feed enough rope through the window to enable the firefighters inside to either make the knot or to connect it to the victim's harness. If a Stokes basket litter is used, the rope is attached to the cradle sling on the litter.

Step 6:   Once the victim is secured to the hauling life safety rope, a guideline is secured to either the harness or the area between the chest hitch and the double loop figure of eight. If a Stokes litter is used, it may be necessary to use two guidelines, one at each end. A clove hitch (with an overhand safety knot) is the knot of choice for attaching the guideline(s). Lower the other

Figure 5.67 Carry the rope up the ladder.

Figure 5.68 At windowsill level, thread the rope over the top of the ladder for three to four rungs and then back over the top again.

end of the guideline out the window so that a firefighter on the ground can hold the victim away from the building as he or she is lowered.

Step 7: Two or more firefighters inside the structure lift the victim out the window. The firefighter on the ground, controlling the descent, takes up the slack in the rope (Figure 5.69).

Step 8: The control person at the base of the ladder, with one foot on the bottom rung for support and using a hand-over-hand grasp on the rope, lowers the victim to the ground (Figure 5.70). The firefighter on the guideline stands at the base of the ladder near to the firefighter doing the lowering to prevent the ladder from being accidentally pulled over, which could occur if the guideline is pulled from a position too far to either side. (NOTE: One firefighter at the base of the ladder can

easily do the lowering and still support the ladder, but it is a good policy to have the ladder tied in or well anchored for safety reasons.)

Figure 5.70 Gently lower the victim to the ground.

USING A LADDER FOR ICE RESCUE

When using a ladder for ice rescue, the immediate goal is to keep the victim from sinking and drowning. The second goal is to complete the retrieval of the victim.

> # WARNING
> It is important that firefighters not complicate the problem by taking unnecessary risks and also falling through the ice.

When the ice is intact between the shore and the spot where the person fell into the water, a ladder is particularly well suited to the job. A ladder works well because it provides a means of

Figure 5.69 Prepare the victim to be lowered by positioning him or her on the windowsill.

reaching the victim from a relatively safe location, the victim can get hold of it, and the victim can then be retrieved with it. The pressure on the surface of the ice, caused by drawing the victim from the water, is distributed over the larger area of the ladder being used. Also, this procedure may prevent further ice breakage. If the ladder will not reach from the shore to the victim, a rope is also used. The steps needed to implement this procedure are as follows:

Step 1: Tie a rope to the shore end of the ladder (Figure 5.71).

Step 2: Slide the ladder across the ice to the victim (Figure 5.72). It may be necessary for the firefighters handling the ladder to lie on the ice to achieve additional reach.

CAUTION: If firefighters handling the ladder have to advance onto unstable ice, lifelines are attached to them. Firefighters involved in the rescue should wear appropriate thermal and flotation protection.

Figure 5.72 Slide the ladder across the ice to the victim.

Step 3: The firefighters with the ladder steady and guide it, while another firefighter holds the rope (Figure 5.73).

Figure 5.71 Tie a rope to the shore end of the ladder.

Figure 5.73 Guide the ladder toward the victim.

Step 4:   After the victim grasps the ladder, the firefighter holding the rope pulls both the ladder and the victim toward shore (Figure 5.74). The firefighters handling the ladder assist.

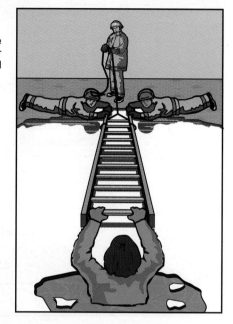

Figure 5.74 Use the rope to pull the ladder and the victim toward shore.

A modification of the procedure for sliding a ladder onto ice to reach a person who has fallen through the ice is applicable for both thin ice and open water rescue. A mounted spare tire or other buoyant object is lashed to one end of a ladder (Figure 5.75). The ladder, with the spare tire attached is slid, tire end first, to the victim. The operation is otherwise the same as that for the ice rescue detailed in the preceding paragraphs.

Figure 5.75 A spare tire may be lashed to a ladder to provide flotation capability.

Fire Fighting Operations

In Chapter 3 of this manual, we discussed proper placement of ground ladders to achieve specific, standard fire fighting tactics. In addition to these "normal" fire fighting uses for ground ladders, there are a number of special ways that ground ladders can be used to support fire fighting operations. The remainder of this chapter highlights some of these special uses.

USING A LADDER FOR VENTILATION

There are several special techniques in which ladders can be used to assist with horizontal ventilation. One method is to use the ladder itself to clear window openings in order to accomplish horizontal ventilation. This technique should only be used in extreme emergencies when ventilation is needed to begin search and rescue procedures. To perform this procedure, the ladder is raised in line with the window, and the tip is dropped against the window glass (Figure 5.76). When performing this procedure, it is best to place the base of the ladder further from the building than would normally be done for achieving a proper climbing angle. This procedure will assure that the tip of the ladder goes completely through the window opening. It also keeps the firefighters at the base of the ladder out of the

Figure 5.76 Windows on upper floors may be broken with the tip of the ladder.

path of any falling glass (Figure 5.77). If the ladder will then be climbed, the base should be moved inward to achieve a proper climbing angle.

CAUTION: Firefighters must be alert for the hazard of falling glass, particularly glass sliding down the beams.

Figure 5.77 When breaking windows, it may be necessary to place the butt of the ladder farther away from the building than it would be if placed for climbing.

If more than one window must be vented, the ladder may be rolled or shifted to the next window, and the process repeated. In this way, many windows can be ventilated in a short time. Keep in mind that this procedure may damage the ladder beams near the top, particularly wood and fiberglass models. Anytime this procedure is used, the ladder must be inspected immediately after use. If any damage is noted, the ladder should be repaired immediately and tested if warranted.

Fire departments who still use negative pressure horizontal ventilation techniques may also find ladders helpful in the placement of smoke ejectors at door or window openings. Smoke ejectors can be hung from or supported on ladders in a variety of ways, including the following:

- In front of window openings (Figure 5.78)
- In door or archway openings (Figure 5.79)
- In stairwells (Figure 5.80)
- Over floor openings (Figure 5.81)
- Over window wells (Figure 5.82)

Figure 5.78 A ladder may be used to support a smoke ejector in a window opening.

Figure 5.79 A ladder may be used to support a smoke ejector in a door opening.

Figure 5.80 A ladder may be used to support a smoke ejector in a stairwell.

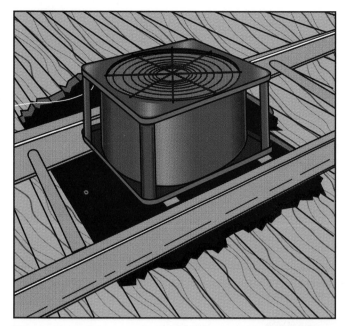

Figure 5.81 A ladder may be used to support a smoke ejector over a floor or roof opening.

Figure 5.82 A ladder may be used to support a smoke ejector over a window well.

Ladders used for this purpose must be inspected for heat damage when the operation is completed. Ladders that show signs of heat damage must be removed from service, tested, and repaired if necessary.

DIRECTING FIRE STREAMS FROM GROUND LADDERS

When a fire stream is going to be directed onto a fire from a ladder, it is necessary to secure the hoseline to the ladder at the vantage point where the operation will take place.

Step 1: Advance a dry hoseline, with nozzle attached, up the ladder to the desired vantage point. Extend the nozzle and approximately two feet (0.6 m) of hose between two rungs so that the hoseline is draped over a rung (Figure 5.83).

CAUTION: The nozzle must be CLOSED during this part of the evolution.

Step 2: Secure a rope hose tool or similar device to the hoseline a rung or more below where the hoseline drapes over the rung (Figure 5.84).

Step 3: Wrap the hook end of a rope hose tool around the rung below the one on which the hoseline is draped. Pass the rope hose tool over the hoseline. If there is still adequate slack, pass it over the rung again on the other side of the hoseline,

Figure 5.83 Drape the hose and nozzle over a rung.

Figure 5.84 Attach a hose strap or webbing to the hose.

Figure 5.87 The firefighter should lock or hook in before operating the fire stream.

and put the hook over the rung on which the hoseline is draped. If there is not enough slack, no second loop is made. Place the hook over the rung on which the hoseline is draped (Figure 5.85).

Step 4: Descend several rungs and have the hoseline charged (Figure 5.86).

Step 5: Assume a position on the ladder to operate the nozzle and either leg lock in or attach a safety belt (Figure 5.87).

Figure 5.85 Fasten the hose strap to the rung.

Figure 5.86 The firefighter should descend a few rungs below the nozzle before the hose is charged.

HOISTING LADDERS

Ladders are sometimes needed on roofs or upper floors of buildings. Hoisting with a rope is probably the easiest and quickest way to accomplish this task. There are two methods, and the one used is a matter of departmental policy.

The first method may be used on any ladder. Its primary advantage is in providing a fulcrum whereby the ladder can be swung into a window or over the cornice or parapet of a roof.

Step 1: Tie a large bowline or figure of eight on a bight knot to form a loop.

Step 2: Place the loop from the underneath side between the rungs of the ladder about one-third the length of the ladder from the tip (Figure 5.88).

Figure 5.88 Thread the loop around a rung that is about one-third of the way down the ladder from the tip.

Step 3: Pull the loop and knot to the end of the ladder, and place the loop over the beam ends (Figure 5.89).

Step 4: Complete the tie by pulling on the standing part of the rope (Figure 5.90).

Step 5: Attach a tag line to the bottom rung of the ladder (Figure 5.91).

Step 6: Firefighters in or on the building take up the slack in the line. The firefighters on the ground assist in lifting the ladder to vertical and guiding it with the tag line.

Step 7: Hoist the ladder up the outside of the building (Figure 5.92).

Figure 5.89 Pull the loop over the tip of the ladder.

Figure 5.91 Attach a tag line to the bottom rung.

Figure 5.90 Tighten the loop around the ladder.

Figure 5.92 Once all the rope lines are attached, hoist the ladder.

As a result of the rope loop being fed from underneath the ladder, the standing part of the rope and the bowline are between the ladder and the building as the ladder is hoisted. This procedure tilts the tip outward as the ladder is hoisted so that it will not snag on windowsills, etc. This arrangement, plus tying it one-third of the way down the ladder, projects the tip above the roof edge or windowsill when it reaches the point where the hoisting firefighters are located. This situation provides a short length of ladder that can be grasped for leverage when the ladder is pulled over the roof edge or windowsill.

The same tie can be used to lower a ladder except that the standing part of the line should be outside the ladder. The procedure swings the butt outward to keep it from catching on obstructions.

A second method of hoisting a ladder is as follows:

Step 1: Place the ladder on the ground on one beam. Thread the end of the rope between two rungs about one-third the length of the ladder from the tip (Figure 5.93).

Step 2: Carry the end of the rope to the butt of the ladder. Tie a clove hitch and a safety around the upper beam (Figure 5.94).

Figure 5.93 Run the rope under a rung about one-third of the way down the ladder from the tip.

Figure 5.94 Tie a clove hitch near the butt.

Figure 5.96 Personnel on the ground may assist with the initial stage of the hoist.

Figure 5.97 The ladder hoist may be completed by personnel on the ropes.

Step 3: Attach a tag line near the butt or to the bottom rung (Figure 5.95).

Step 4: As the firefighters above pull on the standing part of the line, the firefighters on the ground guide the ladder into a hoisting position (Figure 5.96).

Step 5: Raise the ladder to the needed location (Figure 5.97).

Figure 5.95 Attach a tag line near the butt.

BRIDGING FENCES AND WALLS

Bridging fences and walls requires either two ladders or an extension ladder that has been disassembled to make two single ladders. A standard pumper complement of a 24-foot (8 m) extension ladder and a 14-foot (4.3 m) roof ladder is frequently used for this purpose. If the apparatus has two single ladders, this operation is made even simpler.

When the specific point of bridging is not critical, a point adjacent to a fence post is recommended. The procedure is as follows:

Step 1: Raise the larger or heavier ladder, if there is one, into place against the fence (Figure 5.98).

Figure 5.98 Raise the larger of the two ladders against the fence.

Step 2: One firefighter climbs this ladder and leg locks in; the second firefighter passes the other ladder, butt first, to the firefighter on the first ladder (Figure 5.99).

NOTE: If only one firefighter is available, the second ladder is raised, butt first, beside the first ladder before the firefighter ascends it. If three firefighters are available, one should heel the first ladder that is raised.

Figure 5.99 With the firefighter on the ladder locked or hooked in, the butt of the smaller ladder is passed to the firefighter.

Step 3: The firefighter on the ladder grasps the second ladder and slides or passes it over the top of the fence or wall (Figure 5.100).

Step 4: Lower the second ladder butt to the ground. Shift the ladder so that one of its beams abuts a beam of the first ladder; then lash the two together (Figure 5.101).

When low fences or barriers are encountered, the combination ladder can be used. It is opened into the A-frame configuration and then picked up and placed over the fence or wall.

Figure 5.100 Lower the second ladder over the fence.

Figure 5.101 Lash the two ladders together.

SUPPORTING WATER SUPPLY OPERATIONS

There are several ways that ladders can be used to assist in maintaining a constant water supply for pumping apparatus. These include keeping a suction strainer off the bottom, damming the flow of a stream to make it deep enough for drafting, and constructing a catch basin.

Keeping A Suction Strainer Off The Bottom

When drafting water, it is important to maintain 24 inches (600 mm) between the openings of the strainer and the bottom (Figure 5.102). Either a single or a roof ladder can be used for this purpose. The suction hose with strainer attached is brought through the two bottom rungs, or through the second and third rungs from the bottom, depending on the steepness of the bank. This procedure tilts the strainer toward horizontal and keeps it off the bottom (Figure 5.103).

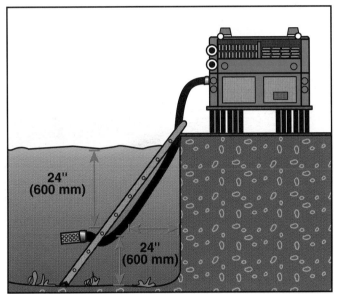

Figure 5.102 There should be at least 24 inches (600 mm) of water all around the ladder.

Figure 5.103 A ladder may be used to keep the strainer off the bottom.

Making A Dam Across A Stream

Sometimes streams are fast running but too shallow for drafting. When a floating dock strainer is available, a ladder and a salvage cover can be used to dam the stream and raise the water level to permit drafting.

Step 1:   Spread a salvage cover on the ground. Place the ladder on one of the long sides (Figure 5.104).

Step 2:   Roll up the ladder in the cover. Leave about 4 feet (1.2 m) of cover free to form a flap (Figure 5.105).

Figure 5.104 The ladder is placed on the edge of a salvage cover.

Figure 5.105 Roll the ladder in the cover until only a 4-foot (1.2 m) flap remains.

Step 3:   Place the ladder and cover assembly across the stream, preferably at a point where the stream bottom is level (Figure 5.106).

Step 4:   Stretch the flap upstream and anchor with rocks or with straight bars stuck through the grommets (Figure 5.107).

It is sometimes necessary to do some tucking at the ends to prevent serious leakage. It may be necessary to support the ladder near its midpoint to prevent the weight of water buildup from bowing the ladder (Figure 5.108).

Figure 5.106 Place the ladder and cover in the stream.

Figure 5.107 Spread out the flap and anchor it with large rocks.

Figure 5.108 Support the center of the ladder to prevent bowing.

Constructing A Catch Basin

A triangular-shaped catch basin is made by lashing two ladders and a pike pole together to form a framework (Figure 5.109). It is then lined with a salvage cover so that it will hold water (Figure 5.110). Catch basins have several uses:

Figure 5.109 Ladders and pike poles are lashed together to form the structure for the catch basin.

Figure 5.110 The catch basin structure is lined with a salvage cover.

- They provide a way for a pumper to get water from a hydrant having different threads than those of the hose fittings carried by the pumper.

- They permit use of hydrants when damaged hydrant outlets prevent the pumper from hooking up.

- They may be used in lieu of a folding portable tank when rural water shuttle operations are necessary.

When a catch basin is used at a hydrant, it is placed in front of one of the hydrant outlets. The side formed by the pike pole should be turned so that it is not in line with the hydrant outlet. The hydrant outlet should be gated so that the flow can be controlled. This procedure is done to prevent overflow and to keep the velocity of the stream of water from causing the salvage cover to come loose from the pike pole (Figure 5.111).

Figure 5.111 The catch basin will hold a few hundred gallons (liters) of water.

When a catch basin is used during a water shuttle operation, care must be taken so that water being dumped does not flow against the pike pole side with any significant force. In all cases, the pumper obtains water by drafting from the catch basin. The pumper's hard suction hose should not be laid across the pike pole side of the catch basin because the weight may bend or break the pike pole.

MAKING A WATER CHUTE

In salvage operations, ladders can be used to construct water chutes for removing water from a building. To do this, one end of the ladder may be supported by either a high piece of furniture or a step ladder. The other end of the ladder should rest on the windowsill. A salvage cover is partially unfolded in order to roll the edges, which are placed inside the ladder beams (Figure 5.112). Note that if the rolled edges are placed underneath the cover, the weight of the water in the trough tends to tighten the rolls. On ladders with high beams, the rolls do not have to be inverted, because the beams will hold them.

Figure 5.112 The salvage cover is placed on the ladder to form the water chute. *Courtesy of Chicago Fire Department.*

Chapter 5 Review

Directions

The following activities are designed to help you comprehend and apply the information in Chapter 5 of **Fire Service Ground Ladders,** Ninth Edition. To receive the maximum learning experience from these activities, it is recommended that you use the following procedure:

1. Read the chapter, underlining or highlighting important terms, topics, and subject matter. Study the photographs and illustrations, and read the captions under each.

2. Review the list of vocabulary words to ensure that you know the chapter-related meaning of each. If you are unsure of the meaning of a vocabulary word, look the word up in the glossary or a dictionary, and then study its context in the chapter.

3. On a separate sheet of paper, complete all assigned or selected application and review activities before checking your answers.

4. After you have finished, check your answers against those on the pages referenced in parentheses.

5. Correct any incorrect answers, and review material that was answered incorrectly.

Vocabulary

Be sure that you know the chapter-related meanings of the following words.

- tactical *(149)*
- precipice *(166)*
- fulcrum *(173)*
- cornice *(173)*
- parapet *(173)*
- draft *(176)*
- velocity *(179)*
- bight *(173)*

Application Of Knowledge

Choose from the following ladder applications for climbing and using ground ladders those that are appropriate to your department and equipment, or ask your training officer to choose appropriate applications. Practice the chosen applications under your training officer's supervision.

- Apply a leg lock on a ground ladder. *(151)*
- Demonstrate one-firefighter roof ladder deployment. *(152)*
- Demonstrate two-firefighter roof ladder deployment (hooks-first method). *(153)*
- Demonstrate two-firefighter roof ladder deployment (butt-first method). *(155)*
- Demonstrate two-firefighter roof ladder deployment (modified butt-first method). *(156)*
- Climb a pompier ladder. *(158)*
- Assist a conscious, physically able victim down a ground ladder. *(160, 161)*
- Bring an unconscious victim down a ground ladder (victim's arms around firefighter's neck). *(161)*
- Bring an unconscious victim down a ground ladder (victim's legs over firefighter's shoulders). *(163)*
- Carry a child down a ground ladder. *(163)*
- Remove an unconscious firefighter leg locked on a ground ladder (unlocking leg method). *(163)*
- Remove an unconscious firefighter leg locked on a ground ladder (cutting rung method). *(164)*
- Lower an extension ladder to a below-grade location. *(165)*

- Use a ladder sling to lower an injured victim from a building. *(167)*
- Use a ground ladder for ice rescue. *(169)*
- Use a ladder and spare tire for ice/open water rescue. *(170)*
- Use a ladder to horizontally ventilate a building. *(170)*
- Use a ladder to support a smoke ejector for horizontal negative pressure ventilation. *(171)*
- Direct a fire stream from a ground ladder. *(172)*
- Hoist a ladder using a bowline or figure of eight on a bight. *(173)*
- Hoist a ladder using a clove hitch. *(174)*
- Ladder bridge a fence or wall. *(175,176)*
- Use a ladder to keep a suction strainer off the bottom of a water source. *(177)*
- Use a ladder and salvage cover to dam a stream. *(177)*
- Use two ladders and a pike pole to construct a catch basin. *(178)*
- Use a ladder and salvage cover to construct a water chute. *(179)*

Review Activities

1. Briefly identify the following:
   - Class I life safety harness (ladder belt) *(150)*
   - leg lock *(151)*
   - horizontal ventilation *(170)*
   - negative pressure horizontal ventilation *(171)*
   - smoke ejector *(171)*
   - bowline *(173)*
   - double loop figure of eight *(167)*
   - figure of eight on a bight *(173)*
   - standing part *(174)*
   - tag line *(174)*
   - clove hitch *(174)*
   - suction strainer *(176)*
   - catch basin *(176)*
   - water shuttle *(178)*

2. Describe proper climbing practices associated with the following key words: *(149-150)*
   - rhythm
   - knees
   - eyes
   - arms
   - hands
   - palms
   - feet
   - tools
   - aerial ladders

3. Explain why one should never use a leg lock on an aerial ladder. *(151)*

4. Explain the main differences between fireground and drill tower pompier ladder use. *(158)*

5. List the order of priority in which victims should be removed at the emergency site. *(159, 160)*

6. Describe general indications of the following rescue priorities: *(160)*
   - highest priority
   - second priority
   - third priority
   - fourth priority

7. Explain what the firefighter should do when a person panics and becomes violent while being assisted down a ladder. *(161)*

8. Explain factors that might necessitate cutting the ladder rung to free an unconscious leg-locked firefighter. *(164)*

9. List the ways in which smoke ejectors can be hung or supported on ladders. *(171)*

10. List uses for catch basins. *(178)*

Questions And Notes

Appendix

_____ **Fire Department**

GROUND LADDER TESTING AND REPAIR RECORD

1. MFGR:	2. MFGR'S Model or Code #:	3. MFGR'S Serial #:	4. FD ID #:

5. Date Purchased:	6. Date Placed in Service:	7. Unit or Location to Which Assigned:

8. Type: ❑ Single ❑ Roof ❑ Extension ❑ Pole ❑ Folding ❑ Combination ❑ Pompier

9. Length:	10. Construction Materials: ❑ Wood ❑ Metal ❑ Fiberglass	11. Beam Type: ❑ Solid ❑ Truss

12. Certified as Meeting NFPA Standard 1931: ❑ Yes ❑ No   Edition Year:

13. Reason for Test: ❑ Annual Service Test ❑ Suspected Damage, Overload, Unusual Use ❑ Exposed to Heat ❑ Retest After Repair

14. Test Date:	15. Person(s) Performing:

16. Heat Sensor Label Check: ❑ Label Unchanged ❑ Label Changed/Heat Exposure Indicated ❑ No Label Present

17. ❑ Horizontal Bending Test Performed   Weight Used: Amount of Deformation: ❑ Passed   ❑ Failed

18. ❑ Hardware Test Performed   Weight Used: ❑ Passed ❑ Failed → Location and Part Failing:

19. ❑ Roof Hook Test Performed   Weight Used: ❑ Passed ❑ Failed → Location and Part Failing:

20. ❑ Pompier Ladder Test Performed   Weight Used: ❑ Passed ❑ Failed → Location and Part Failing:

21. ❑ Hardness Test Performed   ❑ Instrument Calibrated Before Test   ❑ Instrument Calibration Verified Immediately After Test

INSTRUMENT USED: _____   Min. Acceptable Reading for this Instrument: _____

❑ Passed   ❑ Failed-Location of Failure: _____   Failure Reading: _____

22. ❑ Eddy Current Test Performed   Performed By: _____   Firm Name: _____

❑ Passed   ❑ Failed-Location of Failure: _____

23. ❑ Liquid Penetrant Test Performed   Performed By: _____   Firm Name: _____

❑ Passed   ❑ Failed-Location of Failure: _____

24. Status of Ladder as Result of Test: ❑ In Service ❑ Out of Service for Further Testing ❑ Out of Service for Repair ❑ Destroyed ❑ Other

25. Repair Notes: (Date and Initial Entries)
_____
_____

Remarks: (Use Section Number)
_____
_____
_____
_____

_____
Signature of Person Responsible For Test

Glossary

Glossary

Air Mask
See Self-Contained Breathing Apparatus.

Air Pack
See Self-Contained Breathing Apparatus.

Alloy
Substance composed of two or more metals fused together and dissolved in each other when molten.

Angle Of Inclination
Pitch for portable non-self-supporting ground ladders. The preferred angle of inclination is 75½ degrees.

Annealed
Soft state in metal caused by controlled application of heat and cold.

Attic Ladder
Term commonly used for a folding ladder or combination ladder used to access an attic through a scuttle or similar restricted opening.

Baby Bangor
Short-length tapered-truss wood ladder. Also, an attic ladder.

Bangor Ladder
See Pole Ladder.

Base
Lowest, or widest, section of a non-self-supporting extension ladder. Also, the bottom end of any non-self-supporting ground ladder.

Beam
Main structural member of a ladder supporting the rungs or rung blocks. Also called Side Rail or Rail.

Beam Block
See Truss Block.

Bedded Position
Extension ladder with the fly section(s) fully retracted.

Bed Section
Bottom section of an extension ladder.

Butt
(1) Heel (lower end) of a ladder. (2) Act of steadying a ladder that is being climbed.

Butt Spurs
Metal safety plates or spikes attached to the butt end of ground ladder beams.

Certification
Refers to the manufacturer's certification that the ladder has been constructed to meet requirements of NFPA 1931.

Collapsible Ladder
See Folding Ladder.

Combination Ladder
Ladder that can be used as either a single, extension, or A-frame ladder.

Conductor
Substance that transmits electrical or thermal energy.

Continuous Halyard
Halyard whose both ends are attached to the bottom rung of the fly section of an extension ladder. The rope is run from the bottom rung of the fly section, down around the bottom rung of the bed section, and back up to the bottom rung of the fly section.

Corrugated
Formed into ridges or grooves — serrated.

Cracks
Fractures in a material.

Curling
Method for raising a one-firefighter ladder from a flat rest position in preparation for carrying.

D

Deformation
Alteration of form or shape.

Designated Length
Length marked on the ladder.

Dimpled
Depressed or dented (as on a metal surface) to aid in gripping.

Discontinuity
Interruption of the typical structure of a weldment, such as inhomogeneity in the mechanical, metallurgical, or physical characteristics of the material or weldment.

Dogs
See Pawls.

Dry Hoseline
Hoseline without water in it; an uncharged hoseline.

E

Eaves
Lower border of a roof that overhangs the wall.

Electrical Service
Conductor and equipment for delivering energy from the electrical supply system to the wiring system of the premises.

Engulf
To flow over and enclose. In this text, it refers to being enclosed in flames.

Extension Ladder
Sectional ladder of two or more parts that can be extended to various heights.

F

Fascia
Broad flat surface over a storefront or below a cornice.

Fiberglass
Composite material consisting of glass fibers embedded in resin.

Flat Raise
Raising a ladder with the heel of both beams touching the ground.

Flat Roof
Roof that has a pitch not exceeding 20 degrees. A slight pitch is required to facilitate water runoff.

Fly Rope
See Halyard.

Fly Section
Extendable section of ground extension or aerial ladder.

Folding Ladder
Short, collapsible ladder easy to maneuver in tight places such as reaching through openings in attics or lofts.

Footing
Method for securing the base of a ladder.

Foot Pads
Feet mounted on the butt of the ladder by a swivel to facilitate the placement of ladders on hard surfaces.

Front Member
Firefighter working at the front side of a ladder.

Front Of Ladder
Climbing side — the side away from a building.

G

Ground Ladder
Ladders specifically designed for fire service use that are not mechanically or physically attached permanently to fire apparatus and do not require mechanical power from the apparatus for ladder use or operation.

Growth Ring
Layer of wood (as an annual ring) produced during a single period of growth.

Guides
Devices to hold sections of an extension ladder together while allowing free movement.

H

Halyard
Rope used on extension ladders to extend the fly sections. Also called Fly Rope.

Heat Sensor Label
Label affixed to the ladder beam near the tip to provide a warning that the ladder has been subjected to excessive heat.

Heat Treatment
Controlled cooling or quenching of heated metals, usually by immersion in a liquid quenching medium; its purpose is to harden the metal.

Heel
(1) Base or butt end of a ground ladder. (2) To steady a ladder while it is being raised.

Heelman

Firefighter who carries the butt end of the ladder and/or who subsequently heels or secures it from slipping during operations.

Hooks

Curved metal devices installed on the tip end of roof ladders to secure the ladder to the highest point on the roof of a building.

Identification Number

Serial number placed on each ground ladder by the manufacturer.

Inside Hand Or Foot

Hand or foot closest to the ladder or closest to the other member of a two-firefighter team.

Inside Ladder Width

Distance between the inside edge of one beam and the inside edge of the opposite beam.

K

Knee-Foot Lock

Leg position with the knee against the front of a ladder beam and the instep of the foot hooked around the rear of the butt spur on the same beam; used to secure one beam of a ladder while operating the fly section.

Knurled

Having a series of small ridges or beads, as on a metal surface, to aid in gripping.

Ladder

Two rails or beams with steps or rungs spaced at intervals; any fire department ladder of varying length, type, or construction.

Ladder Locks

See Pawls.

Ladder Nesting

Positioning of different width ladders, one partially within another, for storage on apparatus.

Leg Lock

Method of entwining a leg around a ladder rung to free the climber's hands for working while ensuring that the individual cannot fall from the ladder.

Life Belt

Wide, adjustable belt with a snap hook that can be fastened to the rungs of a ladder leaving the hands free for working.

Locking In

See Leg Lock.

Lower In

Procedure for positioning the tip of a ladder against a building after raising.

Lowering

Procedure for removing a ladder from the raised position.

Maximum Extended Length

Total length of an extension ladder with all sections fully extended and pawls engaged.

Mortise

Hole, groove, or slot cut into a wooden ladder beam to receive a rung tenon.

Moving Pivot

Method for positioning a ladder parallel to the objective while raising it.

Nesting

See Ladder Nesting.

Nondestructive Test

Method of testing objects that does not subject them to stress-related damage.

Occupational Safety And Health Administration (OSHA)

Federal agency that develops and enforces standards and regulations for occupational safety in the workplace.

OSHA

Acronym for Occupational Safety and Health Administration.

Outside Hand Or Foot

Hand or foot furthest from the ladder and furthest from the other member of a two-firefighter team.

Outside Width

Dimension from the outside surface of one ladder beam to the outside surface of the opposite ladder beam or the widest point of a ladder including staypoles when provided, whichever is greater.

Parapet

(1) Extension of the exterior walls above the roof. (2) Any required fire walls surrounding or dividing a roof or surrounding roof openings such as light/ventilation shafts.

Park

Rest position of a ladder with one beam resting on the ground and the rungs vertical and perpendicular to the ground.

Pawls

Devices attached to the inside of the beams on fly sections used to hold the fly section in place after it has been extended. Also called Dogs or Ladder Locks.

Permanent Deformation

Deformation remaining in any part of a ladder or its components after all test loads have been removed.

Pitch

Angle between horizontal and a ladder positioned for use.

Pivot

Method for turning a ladder on one beam when the ladder has been raised to a near vertical position.

Pole Ladder

Large extension ladder that requires tormentor poles to steady the ladder as it is raised and lowered. Also called Bangor Ladder.

Protective Clothing

As used in this text: turnout coat, bunker pants and boots, protective hoods, helmet, and gloves.

Pulley

Small, grooved wheel through which the halyard is drawn on an extension ladder.

Rack

(1) Framework used to support ladders while being carried on fire apparatus. (2) Act of placing a ladder on apparatus.

Rail

See Beam.

Raise

Any of several accepted methods of raising and placing ground ladders into service.

Rear Of Ladder

Side closest to the objective — the nonclimbing side.

Rest

Position of a ladder when both beams are resting on and parallel to the ground.

Reverse Curl

Method of returning a one-firefighter ladder to a flat rest position on the ground.

Ridge

Peak or sharp edge along the very top of the roof of a building.

Roof

Outside top covering of a building.

Roof Ladder

Straight ladder with folding hooks at the top end. The hooks anchor the ladder over the roof ridge.

Run Block

See Truss Block.

Rungs

Step portion of a ladder running from beam to beam.

Rungs Away

Position of a raised truss ladder when the rungs are on the side furthest from the objective.

Rungs Down

Position of a truss ladder at rest when the rungs are on the side closest to the ground.

Rung Side

Front or climbing side of a ladder; the side away from the building or objective.

Rungs Up
Position of a truss ladder at rest when the rungs are on the side furthest from the ground.

S

SCBA
Abbreviation for Self-Contained Breathing Apparatus.

Scuttle
Opening in the roof or ceiling providing access to the roof or attic.

Self-Contained Breathing Apparatus
Protective breathing device worn in hazardous atmospheres. Also called Air Mask or Air Pack.

Serrated
Notched or toothed edge.

Set
See Permanent Deformation.

Shoe
Metal plate used at the bottom of heavy timber columns.

Side Rail
See Beam.

Single Ladder
One-section ladder. Also called Straight Ladder.

Splice
To join two ropes or cables by weaving the strands together.

Spotting
Positioning a ladder to reach an object or person.

Staypoles
Poles attached to long extension ladders to assist in raising and steadying the ladder. Some poles are permanently attached, and some are removable. Also called Tormentor Poles.

Stops
Wood or metal pieces that prevent the fly section of a ladder from being extended too far.

Story
Space in a building between two adjacent floor levels or between a floor and a roof.

Straight Ladder
See Single Ladder.

Straps
Strips of webbing with buckles for securing ladders, improvising step ladders, and other tying purposes.

T

Temporary Deformation
Alteration of form or shape that disappears entirely after a load has been removed.

Tenon
Projecting member in a piece of wood or other material for insertion into a mortise to make a joint.

Throw A Ladder
Raise a ladder quickly.

Tie Rods
Metal rods running from one beam to the other.

Tip
Extreme top of a ladder. Also called Top.

Tongue
Rib on the edge of a ladder beam that fits into a corresponding groove or channel attached to the edge of another ladder beam. Its purpose is to hold the two sections together while allowing the sections to move up and down.

Top
See Tip.

Tormentor Poles
See Staypoles.

Torque
Force that produces or tends to produce a twisting or rotational action.

Truss Block
Used to separate the beams of a truss beam ladder. Also called Beam Block or Run Block.

Trusses
Beams consisting of one tensile chord, one compression chord, and truss blocks or spaces between the two.

Tying In
(1) Securing oneself to a ladder; accomplished by using a rope hose tool or belt or by inserting one leg between the rungs. (2) Securing a ladder to a building or object.

V

Ventilation

Systematic removal of heated air, smoke, and/or gases from a structure and replacing them with cooler and/or fresher air to reduce damage and to facilitate fire fighting operations.

Visible Damage

Damage that is clearly evident by visual inspection without recourse to optical measuring devices.

Visual Inspection

Observation without recourse to any optical devices except prescription lenses; may include physical and mechanical examination.

W

Wood Grain

Stratification of wood fibers in a piece of wood.

Working Length

Length of a non-self-supporting ladder measured along the beams from the butt to the point of bearing at the top.

Index

COMMENT SHEET

DATE _____ NAME _____

ADDRESS _____

ORGANIZATION REPRESENTED _____

CHAPTER TITLE _____ NUMBER _____

SECTION/PARAGRAPH/FIGURE _____ PAGE _____

1. Proposal (include proposed wording or identification of wording to be deleted),
   OR PROPOSED FIGURE:

2. Statement of Problem and Substantiation for Proposal:

RETURN TO: IFSTA Editor
Fire Protection Publications
Oklahoma State University
930 N. Willis
Stillwater, OK 74078-8045

SIGNATURE _____

Use this sheet to make any suggestions, recommendations, or comments. We need your input to make the manuals as up to date as possible. Your help is appreciated. Use additional pages if necessary.